AF856387

DYNAMIC INTERNATIONAL OIL MARKETS

Studies in Industrial Organization

Volume 15

Series Editors:
H. W. de Jong, *University of Amsterdam, Amsterdam, The Netherlands*
W. G. Shepherd, *University of Massachusetts, Amherst, U.S.A.*

Advisory Board:
W. Adams, *Michigan State University, East Lansing (Mich.), U.S.A.*
R. E. Caves, *Harvard University, Cambridge (Mass.), U.S.A.*
K. D. George, *University College of Swansea, Singleton Park, U.K.*
E. Heusz, *Friedrich Alexander University, Erlangen-Nürnberg, F.R.G.*
A. P. Jacquemin, *University of Louvain, Louvain, Belgium*, and
Commission of the European Communities, Brussels, Belgium
E. Kantzenbach, *University of Hamburg, Hamburg, F.R.G.*
H. W. Lambers, *Erasmus University, Rotterdam, The Netherlands*
R. Prodi, *University of Bologna, Bologna, Italy*, and
Istituto per la Ricostruzione Industriale, Bologna, Italy
F. M. Scherer, *Harvard University, Cambridge (Mass.), U.S.A.*

The titles published in this series are listed at the end of this volume.

Dynamic International Oil Markets

Oil Market Developments and Structure 1860–1990

by

Coby van der Linde
Leiden University, Leiden, The Netherlands

KLUWER ACADEMIC PUBLISHERS
DORDRECHT / BOSTON / LONDON

Library of Congress Cataloging-in-Publication Data

Linde, Coby van der.
Dynamic international oil markets : oil market developments and structure, 1860-1990 / Coby van der Linde.
p. cm. -- (Studies in industrial organization ; v. 15)
Includes index.
ISBN 0-7923-1478-6 (acid-free paper)
1. Petroleum industry and trade--History. I. Title. II. Series.
HD9560.5.L565 1991
338.2'7282--dc20 91-35263

ISBN 0-7923-1478-6

Published by Kluwer Academic Publishers,
P.O. Box 17, 3300 AA Dordrecht, The Netherlands.

Kluwer Academic Publishers incorporates
the publishing programmes of
D. Reidel, Martinus Nijhoff, Dr W. Junk and MTP Press.

Sold and distributed in the U.S.A. and Canada
by Kluwer Academic Publishers,
101 Philip Drive, Norwell, MA 02061, U.S.A.

In all other countries, sold and distributed
by Kluwer Academic Publishers Group,
P.O. Box 322, 3300 AH Dordrecht, The Netherlands.

Printed on acid-free paper

All Rights Reserved
© 1991 Kluwer Academic Publishers
No part of the material protected by this copyright notice may be reproduced or
utilized in any form or by any means, electronic or mechanical,
including photocopying, recording or by any information storage and
retrieval system, without written permission from the copyright owner.

Printed in the Netherlands

Preface

Writing this book would have been impossible without the help of certain institutions and persons. For a gas-producing and oil-processing country like the Netherlands, there was surprisingly very little, publicly available, research material. Public libraries' collections contained, with a certain degree of inconsistency, little of the more specialised sources. I would therefore like to express my gratitude towards Royal Dutch Shell, and especially the library staff in The Hague, for allowing me to use the company's library, thanking them for their assistance in finding and supplying the required data. I am also grateful for the financial assistance of the 'Nederlandse organisatie voor wetenschappelijk onderzoek' (NWO) and the Faculty of Law of the University of Leiden. They provided the financial means to work a (crucial) month in the very well equipped library of the Oxford Institute for Energy Studies. I am indebted to the staff of the Oxford Institute for Energy Studies, and particularly to Robert Mabro and Jeremy Turk, for their comments, support, and friendship. After I spent a month in the Institute in July 1989, I was able to return for two five-month periods in 1990 and 1991. For both periods, the Oxford Institute and the Leiden Law Faculty provided me with the necessary means.

I would also like to express special gratitude to some people who have been a great support and supplied me with valuable comments at various stages of the study. First, I would like to mention Leigh Hancher for her helpful comments and her editing work. She has been a true friend. I am also indebted to Jacques van Esch and Leo Biegel for their encouragement to embark on this study, and their support to finish this book. Joseph McCahery critically reviewed a draft version of the book, and his suggestions were imperative for its completion. Furthermore, his support in the last stage was invaluable.

I was in the fortunate position to study for my doctorate degree with Prof. dr. H.W. de Jong. His counsel and erudition has impressed me deeply. He created an open and inspiring relationship, which helped me to develop further my ideas on the operation of the international oil industry.

The above mentioned assistance would have been less resourceful without the continuous encouragement, and above all the assurance of profound friendship and love, of my near and dear ones. Without this, I would not have dared to dedicate all my time to this book. They had to endure my antisocial behaviour, my phone calls, and also my impossible single-mindedness. It must have been awful and boring.

Oxford, Summer 1991 — Coby van der Linde

Table of Contents

Tables

Figures

Abbreviations

API	American Petroleum Institute
ARAMCO	Arabian American Oil Company
b/d	barrels per day
BPSR	British Petroleum Statistical Review of World Energy
BTU	British Thermal Unit
cif	cost, insurance, freight
CMEA	Council for Mutual Economic Assistance
CPE	Centrally Planned Economies
EC	European Communities
ECA	European Co-Operation Administration
fob	free-on-board
GCC	Gulf Co-operation Council
GDP	Gross Domestic Product
GNP	Gross National Product
IMF	International Monetary Fund
IPC	Iraq Petroleum Company
kg	kilogramme
KOC	Kuwait Oil Company
LDC	Less Developed Countries
MOICP	Mandatory Oil Import Control Program
mtoe	million tonnes of oil equivalent
NCW	Non-Communist World
NOPEC	Non-Organization of the Petroleum Exporting Countries
o.e.	oil equivalent
OAPEC	Organization of the Arab Petroleum Exporting Countries
OECD	Organization for Economic Co-operation and Development
OPEC	Organization of the Petroleum Exporting Countries
r/p	reserves to production ratio
t.o.e.	tonnes of oil equivalent
UNCTAD	United Nations Conference on Trade and Development
UN	United Nations
US	United States
USSR	Union of Soviet Socialist Republics

Introduction

The analysis of long-term international oil market developments provides considerable insight into the fundamentals of the oil industry's organisation. Such an analysis enables us to comprehend the logic of the pendular movements in the industry's market structures, which have gone from competitive to oligopoly and *vice versa*, and also offers explanation of the industry's periodic crises. The understanding of the logic of changing market structures will also enable an analysis of company behaviour. Furthermore, the oil industry, as one of the basic industries in the economy of our time, has attracted an increasing degree of public and political intervention.[1]

The basic hypothesis of this study, in accord with the thinking of many European economists, is that economic institutions are in continuous motion, because the market coordination mechanisms, competition, coordination and control are never static, but reflect the interaction between the economic process and structure.[2] In this study, the institutions not only include firms (vertically and horizontally integrated, private and state companies) and market structures (competition, oligopoly, monopoly), but also the politically determined

[1] Both types of government intervention involve the redistribution of economic rents and profits. Baumol and Blinder define economic rents as: "any payment made to a factor above the amout necessary to keep that factor in its present employment." The importance of economic rent arises from the fact that it can be taxed away without reducing the quantity of output supplied. The various players in the market, including governments, are rent-seekers. It is important to stress the distinction between economic rents and economic profits. Economic profits are defined as: "the amount a firm earns over and above the payments for all other inputs, including the interest payments for the capital it uses and the opportunity cost of any capital provided by the owners of the firm." Profits can rise above the level of interest rates when: (1) an entrepreneur can establish a monopoly and earn monopoly profits; (2) the income the entrepreneur obtains is a payment for risk-bearing; and (3) the income is a reward for innovation. With regard to profits taxation Baumol and Blinder continue: " So profits in excess of the market rate of interest can be considered as the return on entrepreneurial talent. But this is not really helpful, since no one can say exactly what entrepreneurial talent is. Certainly we cannot measure it;(...) Therefore, we do not know how the observed profit rate relates to the minimum reward necessary to attract entrepreneurial talant into the market-a relationship that is crucial for the contentious issue of profits taxation." W.J. Baumol and A.S. Blinder, *Economics, Principles and Policy*, Harcourt Brace Jovanovich, San Diego/ New York, 1991. The redistribution of economic rents and profits can take place among an economy's interest groups/market players or among one economy and other economies or international interest groups/market players. The distinction between public and political intervention emphasises the different types of market behaviour of governments. Public intervention is executed with the purpose of restoring or seeking internal and external equilibrium, and entails the governments' priority to its economic policy objectives, like balance of payment priorities, stabilising income from exports, price policies, the income distribution in an economy, stable economic growth, etc.. Political intervention is executed to enhance the countries' strategic economic position, and involves the positioning by goverments to gain bargaining power to shift the political and/or strategic balance of power among nations.

[2] In static industrial organization models, change is treated as exogenous, and behaviour and performance are structurally determined. Static models fail to take into account "that competition is an evolving and historic process with possibilities of interactions, going, for instance, from performance to behavior and from behavior to certain structures that thus become endogenous." A. Jacquemin, *The New Industrial Organization, Market Forces and Strategic Behavior*, Clarendon Press, Oxford, 1987, p.3. *See also* H.W. de Jong, European Industrial Organisation: Entrepreneurial economics in an organizational setting, in: H.W. de Jong and W.G. Shepherd (eds.), *Mainstreams in Industrial Organization*, Kluwer Academic Publishers, Dordrecht, 1986.

economic order.[3]

The main assumption is that the interaction between market process and structure results in cyclical movements of the market, but they are not duplicates, because in the course of time, market conditions and structure are never the same. Market process is the pattern of exogenous and endogenous changing market conditions, while market development describes a series of successive market conditions. Market conditions are defined as the market for a certain product or groups of products which is characterized by the combined total of supply and demand factors (like elasticity and cross-elasticity of demand, scale economies, vertical economies, geography of supply and demand, variety in (extraction) costs) at a particular stage or a specific period. In other words, the operation of the balance of demand and supply in (and among) the various stages of production determines the oil market conditions. Market structure is the total of characteristics that, given the market condition, constitutes the position, order and relations of the market participants on the supply and demand side of the market.[4]

The prominence of the time-element in the analysis hinders a uniform definition of the market in terms of product and geographic dimension. Covering nearly 130 years of oil industry development, the markets under study are manifold. The oil products the industry manufactures have changed along with the geographic reach of the market. Moreover, markets do not only expand but also contract, along both dimensions. The crude oil market developed from a local market to a regional, national and international market, while the transitions from one market to another were not always smooth. The various oil product markets have gone through similar developments. In time, we observe oil product markets reduce in terms of volume and geographic reach, like for example kerosene, as a result of technological change and economic development. Consequently, the infrastructure and the organisation of the market required adjustments. Such adjustments take time. The firms' behaviour and performance are also influenced by these shifting markets. They have to incorporate shifting demand and supply conditions in their internal organisation, and adapt their market strategies. This implicit pressure for continuous change runs contrary to the inflexibility of the companies' capital stock and their rate of amortisation, and in stagnating or declining markets enhances collusive behaviour.

This leaves the difficult task of treating the markets under study as consistently as possible, while acknowledging the fact that we are trying to explain something continuously in motion. However, the shifting propensity of markets is exactly what is studied. This leaves no other option than to accept the fact that the market can only be approximated here.

[3] The institutional organisation of economies is political by nature. The inclusion of political dynamics is vital for the understanding of markets. Economic decision-making by governments is based on both economic and political interests. The pursuance of these interests can be on an intergovernmental level (f.i. OECD countries, OPEC, or the Group of 77), on a country level, or by pressure groups within a country. The promotion of political and economic interests can have a large impact on market organisation and the behaviour of other market participants, and can also result in (temporary) alliances among market participants.

[4] H.W. de Jong, *Dynamische markttheorie*, Stenfert Kroese, Leiden, 1985, p.102-104.

The growth-decline or 'European' model, as proposed by H.W. de Jong[5], is based on four theoretical assumptions. First, markets go through evolutionary development stages, such as innovation, expansion or diffusion, maturity and stagnation or decline, depending on the industry in question, and show varying growth rates and other characteristics. Second, industries, especially the larger ones, have a broadening spectrum of individual and specialised products, which prompt product cycles and the rise and/or decline of firms within the industry. Third, changing market structures, reflecting both natural market developments and the strategic moves of the most active players, be it mergers, takeovers, cartel formation, diversification and (inter)regional expansion, etc., follow from the market process, and, in turn, condition the ensuing processes. Fourth, the performance of industry declines in terms of growth rate, added value and similar variables with increasing maturity, unless, which is more exceptional than usual, supply and demand-inspired innovations provide a fresh impetus. These innovations originate frequently outside the industry. As a result, market structures are never static in a longer time-span. Consequently, the cyclical upward and downward movements in processes and structures are held to be more important as the focus of analysis than the given organisational forms. In this study, de Jong's model is extended to capture better the idea that forms of organisation also include the political organisation of markets and economies.

The growth-decline model shows that the market development phases each have their own forms and intensities of competition, which results from the prevailing growth of demand in that phase and the demand elasticities. The growth-decline model features the interaction between market conditions and market structure, while behaviour and performance take less prominence. The power of the growth-decline model is that it makes change endogenous and time becomes an essential factor of analysis. This exacerbates the difference in approach between the growth-decline model and other schools.[6] In contrast, the Anglo-Saxon approach concentrates mainly on the exchanges among structure, behaviour and performance.[7] The problem of this approach is that it does not fully accept the accumulative nature of market determinants from one period to another. Yet, the European and Anglo-Saxon approaches do not exclude each other. Rather, they focus on differing parts of the causation chain, and reflect different time perspectives. This makes the Anglo-Saxon approach less useful for the analysis of a long-term pattern.

Changing market structures (competition, cooperation and control) are, according to de

[5] Ibid.

[6] W.G Shepherd visualises the line of causation in the Anglo-Saxon models in diagrams in his book, *The Economics of Industrial Organization*, Prentice-Hall International Editions, Englewood Cliffs, 1990. It clearly shows the difference in opinion between some European economists and the American approach to Industrial Organization. In the mainstream diagram, the arrow connecting determinants and structure only runs one way, but structure, behaviour and performance have reverse arrows. In the diagrams of the Chicago-UCLA and the contestability model, determinants are in fact not included.

[7] In the Anglo-Saxon approach, the line of causation runs from conditions to structure to behaviour and performance, with some reverse causation possible structure, behaviour and performance.

Jong[8], more adequately analyzed from the perspective of individual economic sectors, thus creating a type of meso-economics, interposed between the micro behaviour of individual firms and the macro-economic developments of large aggregate entities. Hence, the study of a large industry, like the oil industry (in terms of sales the largest in the world), provides an excellent empirical sector in which to employ and evaluate the theory and its basic tenets.

On this approach, the analysis of particular market episodes, say the low-price period of the mid-1960s, or the ups and downs of the 1980s, must necessarily take less prominence. These episodes are not considered of minor importance or of less analytic interest; in fact, they do raise all types of pertinent questions. But it would seem to follow from a long-term analysis, as presented here, that such short-term episodes will not be repeatable, because they are inscribed in the long-term sweep. This, in itself, is an important derivation, because time and again, players and observers of the oil industry have behaved as if they could extrapolate current tendencies, ignoring the underlying fundamentals. In fact these fundamentals are of major importance, and these alone justify a closer study of them.

The oil industry is characterized by counter-cyclical conditions in the upstream and downstream sectors of the industry. Together with the high fixed costs, technical constraints, the asymmetric geographic distribution of oil deposits, the variety of costs among oil wells, the incomplete homogeneity of crude oil, and the inflexibility generated by joint production in the processing chain, these cause the oil industry to be inherently unstable. Periodically, the industry suffers from capacity constraints between the upstream and the downstream sector. These periods are most noticeable when the market enters a new phase of production.

The oil industry developed from a lamp oil producing industry into a fuel oils and petrochemical products producing industry. The periodisation of the 'use' of oil is dissimilar for the seperate consumer markets. These developments, in conjunction with the oil market process, influence the specific oil market structure, the form of organisation of the market participants, and the escalation in governmental intervention. Therefore, the oil market evolution, and in conjunction the methods and the purpose of the participants to regulate the market, is crucial to the understanding of the industry. This suspicion provides another justification for a long-term investigation.

Attention is primarily devoted to the crude oil market. A closer examination of the seperate product cycles in lamp oil, fuel oils and petrochemicals would require seperate studies, which would carry us into fields beyond the scope of the book.

Thus the following chapters will examine the market conditions and the market structures of crude oil production, oil consumption[9] and (international) trade from 1860 to 1990. Market conditions with regard to transport, processing and distribution, technology, the near substitutes for oil and overall (energy) policies of governments are analyzed only in relation to crude oil market developments.

[8] H.W. de Jong, Dynamische Markttheorie, op.cit, p.3-20.

[9] Analysis will be limited to consumption of oil as a fuel and not as a feedstock.

The study is divided into two parts. Part I, comprising three chapters, deals with market conditions and developments between 1860 and 1990; and Part II, also comprising three chapters, discusses market structures after 1945. Chapter one, provides the theoretical framework for analysis of oil market conditions and market structures. In Chapter two, the US market (1860-1940) holds centre stage, because in that period it was virtually the only market. Chapter three broadens the discussion to the international oil industry, which acquired prominence in the period after 1945. In Chapter four, we will look at the impact of the development of the Middle East oil resources by the large oil companies on the international oil market (1928-1970). Chapter five focusses on OPEC's policies, including both its initial successes and the ensuing downfall, which was characterized by the growing lack of cohesion among the member states. In Chapter six, oil market conditions in the 1970s and 1980s will be reviewed with respect to their impact on the market position of the large international oil companies. Chapter seven integrates the findings of Part I and Part II, and the flow of the growth cycle of oil is explained.

Chapter One

Market Conditions, Developments and Structure

1. Theory

A comprehensive analysis of the oil market should cover not only developments in crude oil production and demand, but also changes in the major intermediate stages of production, and in the final product markets. The presence of large vertically integrated companies, with activities embracing the entire chain of production, necessitates the importance of analyzing the market in a broad sense. Moreover, the use of oil is no longer limited to a fuel, but has also developed into an important feedstock for the petrochemical industry. The number of final products derived from oil exceeds a hundred. Demand for oil is therefore influenced in various ways. Its energy use is determined by the fact that oil constitutes a major input in industrial production and in domestic consumption. Its non-energy use depends on developments in other industrial sectors. The evolution of refining and the petrochemical industry, and the continuous extension of the variety of its final products, has generated, to some extent, an ambiguous structure in the production line.

Nevertheless, the oil industry is usually broken down into five distinctive stages; exploration, production, transport, processing and distribution. The economics of extracting a depletable resource are subject to different market conditions, development factors and consequently, a different form of market structure than the manufacture of highly value-added petrochemical products. The number of participants engaged in each stage of production differs, as do other factors such as capital intensity, research efforts, etc. An oligopolistic market structure at one stage does not necessarily entail that this structure is replicated in other stages of production. Varying degrees of competition must therefore be assumed for each of the distinctive stages of the industry. It is, however, impossible to treat developments in the distinct stages of production separately without losing sight of the dynamics in the industry as a whole. The structure of the international oil industry is not necessarily the sum total of the market structures of the separate stages of production, nor are the market structures similar at each stage. Each stage possesses its own characteristic market conditions. Those stages are clearly interconnected.

In the first place, the stages of production behave counter-cyclically. When oil prices are high, exploration for oil increases, production increases, and refining margins are squeezed. The increase in oil production eventually reduces the price of oil, and refinery margins tend to improve. Investments in the various stages of production react to these price incentives, and create the pendular movements of the market between surplus and deficit. This interrelationship of the stages of production as well as the structure of each particular stage itself, pose analytical problems in disentangling the variables at work.

Secondly, many companies in the industry are vertically integrated and involved in international production. They are multinational enterprises. This concept continues to be the subject of new and conflicting types of analysis.[1] There is, as yet, no undisputed theoretical explanation of foreign direct investment[2] and of multinational enterprises, be it Hymer's work on foreign direct investment, Vernon's product cycle theory, Heckscher-Ohlin's trade theory, Aliber's foreign exchange theory or Bain's industrial organisation theory. Moreover, the relevance of these theories to this study appears to be limited as their focus restricts the analysis either to a particular stage of production or to certain market conditions.[3]

Likewise, the extensive literature on the oil industry fails to provide an embracing theoretical framework for the analysis of oil market developments, the pendular movements of the market structures, and the various methods of regulation. The scope of most studies renders valuable but disputed insights in a particular aspect of the industry only. Different theoretical frameworks are used to explain pricing, cost of production, the market conditions of a stage of production and/or the operations of the multinational enterprises, without resulting in a comprehensive analysis of the dynamics of the oil industry. Certain studies on the industry, like Jacoby's and Adelman's[4], although once acknowledged as authoritative, turned out to be limited in scope, because they were overtaken by oil market developments. The main strength of these studies lies in their empirical findings and the analysis of particular periods.

The structure of the oil market remains disputed among researchers with respect to the level of competition in it. According to Jacoby, the international oil industry was effectively competitive from the 1950s on, and the corporate cartel did not actually restrict competition[5]. Others stress the past and present oligopolistic structure of the market and

[1] A summary of the theories is presented by N. Hood and S. Young, *The Economics of the Multinational Enterprise*, Longman, London, 7th impression, 1987.

[2] Foreign direct investment, as opposed to portfolio investment, involves effective control over the investment through the ownership (in part or as a whole) and management of a foreign operation. Foreign direct investment transfers a package of resources (capital, management, technology, distribution outlets, etc.) from one country to another and in addition assumes control over its application.

[3] Theories on the price of oil and competition in the oil industry are summarised (and criticised) by A. Roncaglia in Chapter 3 of *The International Oil Market; A case of Trilateral Oligopoly*, Macmillan, Basingstoke, 1985. Before explaining his own analytical framework of 'trilateral oligopoly' he deals with: the 'user cost' principle developed by Keynes and applied to natural resources by Davidson; Ricardian rent theory; scarcity theory and marginalist theory (Adelman applied this to the oil industry in his theory of decreasing returns); the managerial approach of Penrose; the oligopolist approach of Frankel (he assumed increasing returns to scale); and the 'surplus' approach of Chevalier.

[4] N.H. Jacoby, *Multinational Oil, A study in Industrial Dynamics*, MacMillan Publishing Co./Collier Mac Millan Publishers, New York/London, 1974; M.A. Adelman, *The World Petroleum Market*, Johns Hopkins University Press, Baltimore, 1972.

[5] Ibid., p.5-6. "It is our hypotheses that, when we look at the foreign oil industry in terms of this model, we will be able adequately to measure the vigor of competition and trends in competitive behavior. Viewed in terms of the model, it will be shown that the technical conditions of producing oil in foreign countries, the initial slow

restraints on competition at certain stages of production, which are supposed to restrict competition effectively throughout all stages. The market structure is generally described as an oligopoly with a (varying) competitive fringe.[6]

The extant oligopoly studies have provided a wider understanding of industry structures. According to Moran[7], OPEC is also facing problems common to all oligopolies. The group as a whole can maximise the economic benefits if "the members coordinate production decisions as if they were a monopolistic supplier, but the economic benefits to individual members will be maximized if they can expand their output by offering price discounts without being disciplined by the cartel and without inducing the cartel to fall apart."[8]

Another strand of theory with regard to the oil industry is based on Hotelling's theory of depletable resources, which received renewed attention in the 1970s. Hotelling's theory deals with pricing behaviour of natural resource producers and the determination of marginal costs. The renewed interest was initiated by the emergence of OPEC as a strong international market participant. The increase in oil prices in the 1970s also resulted in renewed studies of oligopolistic/monopolistic pricing behaviour in combination with depletable resource theory. The underlying assumption of the theory of depletable resources is that producers of such a resource will aim at maximizing its present value.[9] Maximizing the present value of a barrel of oil is, however, necessarily different for private international companies, for OPEC as a group, and for its individual member states. This difference stems from the diverging objectives of companies, national states and intergovernmental organizations. The companies needed to maximise the value of their present barrel of oil within the limited time of the concession or production contract, while the oil producing countries have, in principle, a longer time scale over which to maximise the value of their oil.

growth of demand, and the nature of the government controls under which a few American oil companies began foreign oil operations in the 1920s inevitably produced an oligopolistic type of market structure. (...) The rapid evolution of the industry began after the removal of wartime economic controls, between 1945 and 1948. Thereafter, the forces unloosed by the war manifested themselves. The foreign oil industry increased its size by many times, in order to meet burgeoning demand. Hundreds of new firms entered the industry, which assumed an effective competitive structure and behavior."

[6] A. Roncaglia, op.cit., p.4. "The viewpoint of the present analysis is thus to consider the oil market as an instance of multilateral oligopoly. That is, as composed of three groups of economic and political agents, whose behaviour decisively affects the evolution of the oil market: oil companies, producing countries and the consuming countries. (...) Rather, the situation in the oil market could be characterised as a 'trilateral oligopoly': the levels and changes of crude oil prices depend on those factors affecting the degree of oligopolistic control of each group over the market, and on the strategic choices of the 'leaders' within each group."

[7] T.H. Moran, *Oil prices and the Future of OPEC, The political economy of tension and stability in the Organization of Petroleum Exporting Countries*, Research Paper R-8, Resources for the Future, Washington, D.C., 1978.

[8] Ibid., p.iii. "When as in the case of OPEC, the demand for the producers group's output is relatively inelastic while the demand for any individual producer's output is highly elastic, and the marginal production costs are small compared to the group's asking price, the rewards for cheating are great. As a consequence, to maintain the oligopoly, each member must exercise self-discipline in the common good and be assured that his fellow members will do likewise."

[9] J.M. Griffin and D.J. Teece, Introduction, in: J.M. Griffin and D.J. Teece, *OPEC Behavior and World Oil Prices*, George Allen & Unwin, London/Boston/Sydney, 1982, p.14. For a sophisticated discussion of the Prisoner's dilemma involved in oligopoly pricing *see* E. Rasmusen, *Games and Information: An Introduction to Game Theory*, Basil Blackwell, Oxford, 1989, p.21-29.

Griffin and Teece demonstrated the shortcomings of the Hotelling approach. They argued that the idea that OPEC producers priced their oil above marginal cost in the 1970s and early 1980s was wrongly interpreted as an indication of wealth-maximizing monopolistic behaviour.[10] The concept of marginal costs is based on renewable inputs in the production process, and the decision to produce today does not affect the costs of producing in the future. Marginal costs then entirely consist of marginal production costs; capital, labour and material costs to produce the last unit of output. However, according to Griffin and Teece, production of a depletable resource does affect the costs of production in the future and, therefore, also the current value of a barrel of oil produced today instead of in the future. Marginal costs in this case are not only constituted by marginal production costs but also by user costs.[11] User costs (or opportunity costs) depend on future supply and demand conditions. The future demand and supply conditions will vary for each company or for each OPEC member state.[12] The future conditions of supply are dependent on the proven reserves, the technical production capacity, technological advances to improve extraction from a single field, and new supply from other countries. The future demand conditions are dependent on the long term price elasticity[13], income elasticity of oil, economic growth, and the impact of a backstop fuel.[14] A backstop fuel is a fuel that is (economically and

[10] Ibid., p.14 "Knowing that price equals marginal cost in a competitive market, one might be tempted to reason that OPEC is a wealth-maximizing monopolist based purely on the observation that the market price grossly exceeds the marginal cost of producing crude in the OPEC countries ($.10 to $.25 per barrel in most Persian Gulf countries). This fallacy arises from an inadequate understanding of marginal cost."

[11] Ibid., p.14-15. User costs equate to Marginal Revenue in period t minus Marginal Cost in period t [1]: $U_t = MR_t - MC_t$. Since U_t represents the opportunity value of selling the last barrel in period t, the producer may elect to switch production to some other period t' where user costs are higher. In fact, if the producer maximizes its long run profit, it should be indifferent between producing the last barrel now or at any future time period. This result in:[2] $U_0 = U_1 = U_2 = \ldots = U_t$. This equation does overlook the fact, however, that a dollar today is worth more than a dollar in the future. The future User costs should be discounted at the rate of discount r. [3] $U_0 = U_1/1+r = U_2/(1+r)^2 = \ldots\ldots = U_t/(1+r)^t$. The User costs must rise by the rate of discount if the net present value is maximized:[4] $U_1 = U_0(1+r)$.

[12] "While useful in sketching a general path of aggregate self-interest, the rational monopolist approach suffers from representing the OPEC Cartel as if the members are motivated to behave over time as a single unified factor. In reality, the individual governments of OPEC have differing 'rational' economic interests depending on social pressures, revenue needs, alternative sources of exports earnings and fiscal income, hard currency financial assets, and geological reserves. Hence, they have different discount rates for present versus future earnings and different strains or pains associated with holding sparce capacity or not developing additional capacity. Ultimately the members of OPEC have different preferred price and production paths for the exploitation of their petroleum reserves." T.H. Moran, Modelling OPEC behavior: economic and political alternatives, in: *International Organization*, Spring 1981, volume 35, no. 2, p.244.

[13] In Hotelling's 1931 paper, a case was considered of a monopolist facing a demand curve that had a constant elasticity at every point and demonstrated that a monopolist's price path is identical to that of a competitive market. According to Teece and Griffin this is erroneous on two grounds: "First, the demand for oil is not a static function with constant elasticity. It is an empirical fact that the long run demand schedule is much more price elastic than its short run counterpart. Since it may take ten to twenty years to achieve long run adjustment, a monopolist can exploit the short run inelasticity of oil demand by charging an initially high price. A second and a more fundamental objection is that Hotelling's model assumes a fixed and homogeneous stock of oil reserves available at zero cost. When one recognizes that the stock of oil reserves is expansible at higher prices as submarginal reserves become supramarginal, monopoly and competition solutions will differ dramatically." J.M. Griffin and D.J. Teece, op.cit., p.17-18.

[14] "The situation --of monopoly pricing-- is slightly more complex, however, in the case of an exhaustible resource like oil. The existence of a fixed stock of a commodity has an important independent impact on the structure of the monopoly pricing decision. Any source derives its value from the prospect of being extracted and sold. If left in the ground, however, a nonrenewable resource can also generate an implicit return for its

technically) able to replace oil as the main fuel in the future (like crude oil produced from tar sands, oil shales and coal) if prices rise. The pricing goals will be influenced by the expected supply of a backstop fuel at a certain date in the future. The wide diversity of the reserve-to-production (r/p) ratios of the OPEC member states (See Table 5.2) is enough to result in different discount rates and subsequently in a desired current price for oil. The same reasoning applies to oil companies. The varying rate of vertical integration, the divergent companies' oil reserves, and concession expiration dates will result in different discount rates.

Accordingly, the diversity of the OPEC member states will also be expressed in their discount rates and therefore their national optimal pricing goals. Collectively, OPEC can strive for wealth-maximization, but it cannot do so for its member states. The fact that OPEC is an intergovernmental organization, and the member states continue to enjoy sovereignty over their national petroleum policies, combined with the differences in optimal pricing goals, prevents the pursuit of a wealth-maximizing policy by OPEC. Indeed, OPEC's policy has been the outcome of national income goals and national long-term pricing goals. Since 1973, OPEC's market behaviour has had elements of a price-fixing, a market-sharing and a target revenue cartel.

The income needs of the member states are determined by spending needs and plans. Governments that try to administer pricing and production with regard to target incomes face the continuous shift in 'income requirements'. The member states will, therefore, have to reach a compromise with other OPEC member states which might hamper the achievement of national pricing and production goals. Furthermore, OPEC has to cope with the growing disparity in the member states' income requirements. The limitations to growth of the absorption capacity of some member states has impeded OPEC's ability to regulate prices and production satisfactorily for all member states.

Notwithstanding the valuable theoretical and empirical insights offered by these economic analyses, for instance, their theoretic studies on price behaviour or a particular market structure, like oligopoly, they fail to provide a comprehensive framework with which to analyse the flux of the oil market and the resultant oscillations in market structures. The analyses are often restricted in time, to a certain phase of production or to a particular market structure or a combination of the above. By omitting the pendular movements of market as a central concept, these theorists' explanations of these structural changes are restricted to exceptional circumstances, or as a result of (political) government intervention. Market behaviour of governments, important with regard to the development and the structure of the oil market, should, like the market behaviour of other market participants, be explained within the context of the dialectics of market coordination mechanisms, which

owners by appreciating in value. Thus to the analyses of elasticities of supply and demand, and the determination of a suitable discount rate, the rational monopolist in the oil industry must add considerations about the art of exploitation that balance oil in the ground and money in the bank." T.H. Moran, in: *International Organization*, volume 35, no. 2, p.242.

are an expression of the relationship between the economic process and the form of organisation. Therefore, the growth-decline paradigm will be discussed next.

2. The growth-decline paradigm

De Jong has conceptualised the growth-decline paradigm in his book *Dynamische Markttheorie*.[15] There, he explains the relationship between changes in the sectoral structure and the development of a product by means of the growth cycle model.[16] This concept entails structural tendencies in supply and demand, and focusses "on the various stages through which products, processes, and sometimes whole industries pass from birth to death".[17] The product cycle concept is not viewed as deterministic, or as a simple mechanical law of evolution. The product's life cycle is continuously modified by the introduction of new products, new systems of production and organisation.[18] This breaks repetitive processes and provokes mutations. The coordination mechanisms adapt themselves to the phase of market development.[19]

The (theoretical) growth cycle can be divided into four phases or stages: (1) introduction or innovation phase; (2) expansion or imitation phase; (3) maturity phase; and (4) stagnation phase. The presentation of the theoretical growth cycle has no implications for the reality of any given product cycle with respect to the sequence of the stages, their form or duration, and it is not necessary, especially in a large cycle, to connect the structural changes to the entire product cycle. Growth cycles can be observed in products of technical origin or natural products that are processed by means of modern production methods.

In the course of the growth cycle, three essential moments with regard to changes in the market structure can be recognized. The first essential moment is when a process of deconcentration begins to take place. It occurs in the transition between the introduction phase and the expansion phase, and this process continues during a large part of the latter, when the innovating firm will lose its monopolistic advantage over others and competition will increase. When the potential market is large in relation to actual sales, the entrepreneurs, and new entrants, have an incentive to build up new capacity. This new capacity can, depending on transport costs, be achieved by building more plants of the same size or building larger plants which capture economies of scale. Under these circumstances there will be a tendency towards deconcentration. The level of growth of the market enables more numerous, independent suppliers to reach higher levels of optimum output, and existing firms

[15] H.W. de Jong, 1985.

[16] See Introduction.

[17] H.W. de Jong (ed.), Market structures in the European Community, in: H.W. de Jong (ed.), *The Structure of European Industry*, second revised edition, Kluwer Academic Publishers, Dordrecht/Boston/London, 1988, p.16.

[18] A. Jacquemin, op.cit., p.196.

[19] H.W. de Jong, op.cit., in: H.W. de Jong (ed.), *The Structure of European Industry*, 1988, p.14.

or new entrants will be inclined to focus initially on one phase of production in order to release or target the scarce resources which are necessary for expansion. The level of competition in the expansion phase depends not only on the course of the diffusion process[20] but also on the new entrants' technological improvements to the production process. In the oil industry, technological advances in both the upstream and downstream phase of the production process have been a major influence on market developments.[21] Drilling techniques, means of transportation (pipelines, tankers, etc), refining techniques and the development of the petrochemical industry have not only had an impact on the market structure in that particular stage of production, but also on the structure of the industry as a whole.[22]

A second essential moment with regard to the change in market structure occurs when both vertical and horizontal concentration increase during the transition from the expansion phase to the maturity phase, as well as during the latter. Price competition loses its effectiveness in driving competitors out of the market if the prospect of large cost decreases is absent and the product has acquired a low price elasticity of demand.

A third essential moment occurs during the maturity phase when diversification increases to avoid the uncertain outcome of price competition on profits and as an alternative means to improve market share. However, diversification of products has little importance for raw material markets and semi-finished products. In order to avoid price competition, producers will either be inclined to conclude cooperative agreements (cartels or syndicates), or will merge or take over other firms, or will be encouraged to establish a price leadership. The manner of collusion is dependent on several factors. Competition policy can prevent the conclusion of cartels. Mergers or takeovers can be favoured where they lead to large economies of scale. The presence of a few strong, independent, firms facilitates coordinated market behaviour. The presence of state oil companies (politically) restricts merger or take-over opportunities, and other forms of collusion will be pursued. Presently, the number of joint-ventures and other contractual forms of cooperation in the oil industry between private and state oil companies is increasing. The emergence of new forms of cooperation between market participants can be explained in various ways. However, the political factor, especially in developing countries, is important. The fact that OPEC countries gained control over their natural resources, which was generally welcomed by developing countries as the beginning of a new international economic order, makes it politically undesirable to return

[20] The course of the diffusion process is dependent on the long-term advantage of the new product in relation to the existing products and production processes, the costs incurred by the adoption of the product and the insecurity resulting from the new product or production process. The length of the introduction phase and the moment of transition to the expansion phase is determined by these factors. H.W. de Jong, Dynamische Markttheorie, p.134-135.

[21] See section 5 and 6.

[22] See for more details on the early history of the oil industry and the impact of new technology and competition, for instance: C. Tugendhat and A. Hamilton, *Oil the biggest business*, Eyre Methuen, London, 1975; *See also* H.F. Williamson, R.L. Andreano, A.R. Daum, G.C. Klose, *The American Petroleum Industry, the age of energy 1899-1959*, Northwestern University Press, Evanston, 1963.

to some sort of concessionary system to develop these resources. The fact that some countries lack the investment capital, skills or both to develop their resources, brought about these new forms of cooperation.

The model further contends that horizontal and vertical concentration occurs during the introduction phase, while the expansion phase is characterized by horizontal and vertical deconcentration. Vertical integration increases again in the maturity phase and diminishes in an advanced stage of the stagnation phase, while horizontal integration increases from the end of the expansion phase into the stagnation phase.[23] Diversification of activities by a mature company, which is confronted with temporary or permanent stagnation in its main sectoral activities, and which initiates take-overs of innovative companies still in their expansion phase, must be seen as a sign that the integrated company is advancing further into the maturity phase or entering the stagnation phase of production.

Integration is, among other things, the result of transaction costs.[24] The organisation of production in a firm saves certain market costs, assuming that factors of production can be obtained at a lower cost than the market transactions, which are superseded. If the organisation of production in a firm fails to reduce the costs of production, it is always possible to return to the market.[25] The likelihood of vertical integration is explained in terms of the size of the transaction costs, the specificy of investments, the complexity of inputs, risks and market size.[26] In addition, de Jong argues that the urge to eliminate transaction costs is most strong when monopoloid market structures (for instance as a result of economies of scale or scope) exist in the parts to be integrated or when dynamic factors are involved.[27] Concentration of the market can vary, because the size of the firm is limited as a result of diminishing returns of management, changes in the level of uncertainty and risk, and/or the growth or contraction of a market. In the perspective of the growth-decline model, the margin, at which the costs of organising within the firm is equal to the organisation in another firm or leaving the transaction to be organised by the price mechanism, must, therefore, be considered as a function of the phase of production in a

[23] H.W. de Jong, Dynamische Markttheorie, p.105-116. A graphical presentation of horizontal integration and deconcentration can be found on page 108, Figure 6 and of vertical integration and disintegration on page 112, Figure 7.

[24] Transaction costs are defined as search and information costs, bargaining and decision costs, policing and enforcement costs. R.H. Coase, *The Firm, The Market, and The Law*, The University of Chicago Press, Chicago/London, 1988, p.6 With regard to transaction costs and concentration Coase states: "In my article on 'The Nature of the Firm' I argued that, although production could be carried out in a completely decentralized way by means of contracts betweeen individuals, the fact that it costs something to enter into these transactions means that firms will emerge to organize what would otherwise be market transactions whenever their costs were less than the costs of carrying out the transactions through the market. The limit to the size of the firm is set where its costs of organizing a transaction becomes equal to the cost of carrying it out through the market. This determines what the firm buys, produces, and sells." Ibid., p.7.

[25] Ibid., p.40.

[26] O.E. Williamson, *The Economic Institutions of Capitalism*, The Free Press, New York, 1985. *See also* J. Tirole, *The Theory of Industrial Organization*, The MIT Press, Cambridge, Massachusetts, London, 1989, p.33-34.

[27] H.W. de Jong, Dynamische Markttheorie, p. 187.

growth cycle.

3. Oil industry growth cycles

The growth-decline model mainly addresses a whole industry or distinct parts of the processing chain. However, the model is partial because it is unable to explain the interaction between the different parts of the processing chain. In terms of the international oil industry, the level of aggregation of the model ignores the pivotal relationship between the upstream and downstream sectors of the oil industry. The counter-cyclical developments of these sectors, and the effect on the market process, necessitates further disaggregation of determinants in order to extricate the fundamentals of the industry. Consequently, it is exactly the logic of the separate parts of the production chain and the way they interact that requires disaggregation of the growth cycle into its composites.

The term 'growth cycle'/'product cycle' has not always been carefully used. The concepts are (interchangeably) used to describe both the evolution of process and product. These two concepts should be clearly separated in order to avoid confusion, particularly when an industry exhibits more cycles, each with its own logic, each developing in its own time and space. For this purpose, four types of cycles will be distinguished that cover different dimensions of the oil market: (1) an industry growth cycle; (2) a crude oil cycle; (3) an oil product cycle; and (4) a market cycle.[28] This amplification is required to extricate underlying fundamentals in the oil industry, which are not captured by the more traditional distinctions. The industry growth cycle refers to the combined outcome of crude and oil product cycles and the market cycle. This cycle conforms to the interpretation of the growth-decline paradigm. The crude cycle reflects the development of crude production and consumption, which is, among others, determined by the composition of the various oil product cycles and reflects the dominant 'use' of oil. An oil product cycle refers to the evolution of a particular (oil) product, like fuel oil, kerosene, gasoline, etc.. The particulars of the crude and oil product cycle are not further addressed here, since they largely correspond to the description of the (industry) growth cycle. The market cycle addresses the supply and demand relationship at a given price level in the course of a growth cycle, i.e. a buyers' or a sellers' market, and its effects on capacity investments in the upstream and downstream sector of the industry. It is this particular cycle that captures the interaction between the upstream and downstream sectors, which makes it the one of particular interest.

Contrary to the other cycles, the market cycle should not be seen as a S-shaped curve, nor as a singular line. This cycle embodies the interaction between two, in principal, counter-cyclical reactions to the same market conditions, and is best pictured as a chain, with uneven

[28] These first three types of cycles are, of course, all 'growth' cycles. The name given to the crude cycle, industry growth cycle, which encompasses the oil product cycles and the market cycles, might be confusing. However, this name fits its description best. Such a cycle reflects the longer supply and demand developments in the product markets.

rings, in the space of both the crude oil cycle and the separate oil product cycles.

In a crude oil sellers' market, producers have a strong strategic market position because buyers want to buy more at a given price level than the sellers can or want to sell or produce. In such a market, stocks will diminish, prices and profits of crude producers will have a tendency to increase, while those of crude oil buyers tend to decrease. Crude oil producers will be stimulated to increase capacity, and buyers will tend to integrate backward and rationalise processing, intensify investments, and apply innovations. New entrants are attracted into the crude oil market.

In a crude oil buyers' market, crude oil producers can or want to sell more than buyers are prepared to buy at a given price level. In such a market, stocks increase, prices and profits decline, while profits of crude oil buyers increase. Crude oil production tends to diminish, and capacity becomes less utilised. Depending on the duration of the buyers' market, crude oil production capacity diminishes in the short term, when wells have to be capped. Crude oil producers will move along the (vertical) production line, while oil processors (buyers) are stimulated to increase capacity. Competition among crude oil producers is strong.

The price elasticity of both crude and oil products' demand has a tendency to decrease from a very high level in the introduction phase to a low level in the maturity and stagnation phases, when the market becomes increasingly saturated. In the expansion phase, additional supply from new entrants and established firms can easily be cleared in the market at a lower price level. The innovations in the production process and the economies of scale enable the producers to increase turnover and/or to increase their profit margins. In this phase, the income elasticity of demand is also fairly large. Contraction of economic activity and demand in this phase will increase price competition among the firms, which diminishes excess capacity. At a given price level, however, the expansion of demand subsides over the course of time, while supply increases. The demand and supply curves increasingly intersect at a wider angle. The shift in the supply curve reflects innovation, increased factor productivity, and economies of scale. The rightward moving and sloping supply curve indicates that at a given price level excess capacity increases.[29] This means that the market can only be cleared at continuously lower price levels. Since production costs differ, low cost producers will increase supply to maintain their market share and profits. The increasing excess capacity reduces, in turn, the influence of changes of demand on the price level.

In principle a sellers' market corresponds with an expansionary phase, and a buyers' market with a declining phase of the growth cycle, but obviously there is no necessary strict connection. During the maturity and stagnation phases both sellers' and buyers' markets occur, and during these phases the market cycle changes more often. The succession of these cycles describes the long term development of a market.

A sellers' market can arise as a result of a boom period, when demand increases rather

[29] H.W. de Jong, *Dynamische Concentratietheorie*, Stenfert Kroese, Leiden, 1972, p.86.

suddenly. The oil industry might not be able to meet demand (for specific products) because the increase in supply is subject to delays or is technically restrained. These capacity bottlenecks will create price spikes. Such a market can also arise when producers constrain supply, or when supply is suddenly (partially) interrupted. A buyers' market can arise when demand suddenly slumps as a result of a recession, substitution, or growth of demand is overestimated by producers and/or a sudden large increase of supply, for instance when a large oil field is brought on stream. However, these are only partial explanations of swings in the market cycle. The more fundamental underlying investment patterns in the upstream and downstream sectors provide a more profound explanation of these swings.

Upstream and downstream capacity investments are, for the most part, sensitive to current price levels of crude oil. The fact that the investments come on stream with a time-lag causes capacity continuously to either under- or over-shoot the equilibrium capacity level. Therefore, investments patterns in the current period determine the swings in the market cycle in the next period.[30]

The response of the upstream and downstream sectors to the market conditions are allegedly counter-cyclical. However, this is not always the case. The irregularity of the response is partly due to particular characteristics of the industry itself, and partly due to government intervention. On the supply side, rising prices stimulates exploration expenditure, while declining prices causes investments to contract. The upstream investment decisions are also influenced by the present utilisation rate of capacity in place, and demand expectations. The capital intensity of exploration, the fact that investments are made without absolute certainty about the quantity and quality of oil that can be recovered, and the time-lag between decision and production increase the investment risk. The time-lag between decision and production can bring the new oil production on the market under rather different conditions. However, once production capacity is in place, oil will be produced from that well as long as it covers the low operating costs, thus aggravating the (excess) capacity bottleneck. On the demand side, the responsiveness of demand to price changes becomes less in the course of the growth cycle. Therefore, the market can only increasingly be cleared at lower price levels. Exploration activity declines and oil production will approach or exceed the replacement ratio of reserves until new capacity bottlenecks arise and the price trend turns.

The reaction of the downstream industry to crude oil price trends is basically opposite, since oil price fluctuations influence the processing margins in a reverse way. Investments in capacity extensions increase when prices of crude oil are low compared to the processing operating costs. The time-lag between investment decision and actual operation is, like exploration, large, and capacity can come on stream under different market conditions. This type of investment usually occurs in the expansion phase. In the introduction or maturity

[30] The unstable nature of the market influences market behaviour and performance of the various market participants. For instance, oil companies, unless they are able to adopt a long-term perspective on the market, will display a pro-cyclical investment behaviour.

phase, rising crude prices stimulate manufacturers to move upward on the processing chain towards the less price- and income-elastic oil products. Their decision to deepen and intensify investments structurally changes the capital stock in this sector.

The reaction of the downstream industry is also influenced by the price trends on the product markets. As with crude oil, the price elasticity of oil product demand declines in the course of a growth cycle. However, the evolution of the elasticity of oil product demand varies by product and also with crude oil elasticity developments. Moreover, substitution can be induced by government policies, when other fuels are favoured or subsidised. The divergent (in)elasticities in the product markets and in the crude market, the fact that these products are derived from joint-production, and the capital intensity of processing inhibit quick adjustments to changing market conditions. The counter-cyclical developments between the upstream and downstream sectors are, therefore, not each other's perfect mirror image.

Investments in the various stages of production react to price, inducing diverging capacity investment patterns for the upstream and downstream sectors. The different logic of the upstream and downstream investments is reinforced by the high fixed costs in certain phases of production, technical constraints, the asymmetric geographic distribution of oil deposits, the variety of costs between oil wells, the incomplete homogeneity of crude oil, and the inflexibility generated by joint production in the processing chain. This creates capacity constraints between them, and causes the pendular swings of the market between surplus and deficit. It makes the industry inherently unstable. As a result, the market generates capacity bottlenecks at and between the various stages of production, and the inability of the market participants to prevent these leads from one shock to another.

In particular, these fundamental counter-cyclical developments and the capital intensity of exploration and processing have been an incentive for vertical integration of companies. Vertical integration enabled the companies to internalise part of the fluctuations of the market cycle, reducing their investment risks, and, in general, allowed for a longer-term perspective of the market. It also made it possible to counterbalance, depending on the degree of integration, profits and losses in the different stages of production, reducing taxation and otherwise serving their long-term interests. Other factors influencing forward and backward integration have been geological, geographic, technical and infrastructural constraints. Furthermore, the incomplete, or absence of, flexible markets placed a premium on security of crude supply for producers and processors alike. It gave refiners the desired security of supply, and producers of oil a more secure market outlet.[31]

The oil industry is also increasingly influenced by public and political intervention. The impact of the promotion of national interests by governments on the course of the growth cycles can vary between measures concerning the macro-economic climate in a country, and large political upheavals. The exchange rate policy of a country, the level of government expenditure, the organisation of the banking system, the legal constraints on business

[31] See Chapters two and four.

practises, the cooperation and integration of countries in intergovernmental and supra-national organisation are among those factors that influence the course of the (international) economic process. Examples of political intervention with a significant impact on the oil industry include: the nationalisation of the oil companies' assets after the Russian Revolution; the possibility for France and the United Kingdom (after the US returned to its isolationist policy after World War I) to divide the Middle East among themselves at the San Remo Conference, subsequently securing their oil interests; the emergence of the Non-Alignment Movement in the 1950s as a result of the de-colonisation process, the rise of OPEC, and the subsequent nationalisation of the oil companies' assets; and finally the 1990 Gulf crisis.

The modern history of the oil industry has, so far, brought about three main growth cycles, based on the dominant use of oil: a lamp oil cycle (1860-1900); and two distinct fuel cycles (an American cycle: 1900-1940; and an International cycle: 1950-). These growth cycles are composed of various oil product cycles. The fuel cycles are, for instance, composed of the kerosine, fuel and diesel oil, naphtha, and gasoline cycles. During the first fuel cycle heavy and middle destillates prevailed, while the second fuel cycle was dominated by lighter products. These oil product cycles can vary in terms of type of growth cycle, like new applications, process renovation or repeated cycles. The oil product cycles also have their own introduction, expansion, maturity and stagnation phases, which can vary in extent and duration. Both slope and the duration of the crude oil cycle and the oil product cycles can also be influenced by the market cycles. The combination of market cycles and the various oil product cycles result in the particular slope and duration of the industry growth cycles.

Output of the oil industry has expanded since 1860 with the transition from one main product cycle to another. However, the transition from one phase of a product cycle to another, or from one main growth cycle to another, has not been a smooth and balanced process. Delays in introducing the product or products onto the market, the rate of diffusion of the production process, and the opportunities for product differentiation can, along with other factors[32], influence the extent and duration of certain phases of the growth cycles and can also inhibit the smooth transition from one phase or cycle to another. Moreover, technological progress is irregular and every innovation seems to have its own life-span. The transition of the American oil industry, for instance, from a lamp oil producer to a producer of major fuels between 1880 and 1914 was not smooth, and the growth of consumption and production stagnated temporarily (See Table 2.6).

An industry growth cycle is not only determined by oil product cycles, but also the fusion of national and/or regional markets into a single world market. The internationalization of the oil industry took place in several stages, with the large international oil companies as the main agents. This fusion of markets did not develop consistently and this, in turn, affected the course of the growth cycles. Also governments influenced interfuel competition in various

[32] See section 5 on the particulars of oil and oil products.

ways, protecting domestic produced fuels, and thus inhibiting the full expansion of a particular oil product market. These inconsistencies could dampen expansion or even create temporary stagnation. Another cause for temporary stagnation could be the saturation of one (domestic or product) market before new (foreign or product) markets were developed and/or by swings in the business cycle.

Table 1.1 Composition of oil product demand in percentages

	1970	1980	1990
US			
gasoline	39.3	38.6	41.9
middle dist.	24.6	26.0	29.4
fuel oil	16.5	16.6	8.5
other	19.6	18.6	20.1
West Europe			
gasoline	16.6	20.8	25.0
middle dist.	32.2	34.5	40.1
fuel oil	37.8	31.2	17.7
other	13.4	13.6	16.6
Japan			
gasoline	15.3	17.8	21.4
middle dist.	16.1	24.7	34.7
fuel oil	56.1	43.4	26.6
other	12.5	14.2	17.6
LDC			
gasoline	17.0	18.1	18.6
middle dist.	29.9	32.2	38.7
fuel oil	38.4	35.9	27.2
other	14.6	13.6	15.5

source: BPSR 1980, 1991.

The penetration of oil varied among the consumer markets. The American oil market was the first to expand, and was followed by the expansion of the European and the Japanese markets. The penetration of oil is related to the level of activity of the economy and the style of the production system. In expanding markets, the oil/GNP ratio was above unity. The relationship of oil consumption and economic growth, expressed in the oil/GNP ratio, diminished in the course of time when the production system shifted from energy intensive to more knowledge and capital intensive systems. This shift in economic activity resulted in a change in the pattern of oil consumption. The average annual growth rate of consumption became smaller, and demand shifted to the lighter end of the barrel. The shift in maturing markets to the higher value added (lighter) fuels first occurred in the US, and later in Western Europe (See Table 1.1). Lately, the non-energy use of oil gained more importance in these mature fuel markets. Thus, the maturity of markets is expressed both in quantitative and qualitative terms.

The presentation of product cycles as smooth S-curved lines, depicting the course through the introduction, expansion, maturity and stagnation phases, should be considered as trend lines. In reality, market cycles, originating in business cycles and various other circumstances, alter the market conditions. The market cycles thus influence the course of the product cycle. The industry growth cycle, for instance, also represents the impact of the incomplete fusion of the American oil market with the international oil market in the 1930s[33] and the (temporary) separation of the American market from the world market in 1958. The last was an example of the impact of government intervention which had a profound influence on market conditions, and subsequently on the course of the growth cycle.

The impact of the 'three essential moments' in the course of a growth cycle on the market structure can also be confined to a particular stage of production, and to a particular oil product, and does not necessarily occur across the industry as a whole. Although this also influences the course of the industry life cycle, the dominant market structure for the industry as a whole might absorb these developments or delay their (perceivable) influence. The complexity of the oil industry, producing interrelated products from one (not completely homogeneous) feedstock in diverging sub-markets, the geographic limitations of production, development of consumer markets and government intervention, might well confound an accurate determination of market conditions, developments and market structures.

4. Cartels

In the course of time, the oil market has several times been subject to cartelization. A cartel is a group of firms (or market participants) acting as a single monopolist, and it strategically aims to bar (preempt) new entrants from the market by crowding the product space.[34] The persistence of a cartel or monopoly is according to Tirole limited, because: (1) an entrant can possess superior technology; (2) there is no time for the established firm(s) to preempt entry; and (3) the established firm(s) may not have complete information about the entrant's characteristics.[35] In terms of the growth-decline model, cartels are by definition limited in duration because they are the result of particular market conditions in the course of a growth cycle.

[33] See Chapter 2.

[34] J. Tirole, op.cit., p.346.

[35] Ibid., p.351-352.

The number of definitions of a cartel is substantial.[36] Yet, these definitions do share common characteristics: 1) cartels are defined as agreements and/or non-agreements based on reciprocal behaviour of the economic subjects; 2) the cartel members remain legally/ financially/ organizationally independent economic units; 3) the definitions stress the differences of ends and means of cartels; 4) the restrictions on competition are determined by the ends and means of cartels; 5) generally the definitions only apply to supply cartels.[37] In general, a cartel is defined as: an agreement (formalized or by acquiescence) between independent market participants in the same or parallel sector, which are (potentially) competitors in the same market(s) or good(s), have action parameters in common and attempt to regulate or influence the market conditions to their own benefit. The agreement between the independent market participants can be both a result of force or of free choice, formalized or in unofficial acquiescence. The independent market participants can be private or state companies, or a mixture of entities. The market participants should, however, be potential competitors in specific market(s) or good(s) and therefore be financially and organizationally independent. This definition implies a cooperative game without conflict.[38]

Cartels may take a wide variety of forms, whether they are classified by type or by action parameters. There are price cartels, production cartels, market sharing cartels, syndicates etc. It is more important to establish the probability of cartel formation in a particular market. One prerequisite for cartel formation is a low price elasticity of demand and supply. Another requirement is some sort of cohesion and/or sufficient common interest among the cartel members. A small group of (large) cartel members will constitute a more stable cartel than a larger group of smaller members, but a small number is not a prerequisite. The cartel gains stability if the number of products or sort of products is limited and if the products are homogeneous.

Furthermore, technological innovation should not generate a short lifespan for the product subject to cartelization. The rate of return on new products or production processes needs to be sufficiently high to collect the research and development costs in a relatively short time. Increasing investment costs will increase price competition among existing market participants in the introduction and expansion stage of a product. In the electronic components industry, for example, these increasing costs as a result of a short life-span of the products, deterred new entrants into the market.[39] Although cooperation among market participants in research and development has become common, cartelization seems less likely,

[36] H. Cox, Kartelle- Strukturanalyse, Wettbewerbswirkungen und wettbewerbspolitische Behandlung, in: H. Cox, U. Jens, K. Martert, *Handbuch des Wettbewerbs*, Verlag Franz Vahlen, München, 1981; and *see also* H-H. Barnikel, *Theorie und Praxis der Kartelle*, Wissenschaftliche Buchgesellschaft, Darmstadt, 1972.

[37] H. Cox, op.cit., p.231.

[38] E. Rasmusen, op.cit., p.30.

[39] R. Ballance and S. Sinclair, *Collapse and Survival: Industry Strategies in a Changing World*, Allen and Unwin, Boston/Sydney, 1983, p.146.

at least for an extended period of time.[40] The likelihood of cartel formation is larger at the end of the expansion stage/beginning of maturation stage and in the stagnation stage. In this stage of the product cycle there is an increase in market concentration, per unit costs and a decrease in profitability.[41] Depending on the differences in strengtamong the market participants, a variety of collusive behaviour can materialize, including cartel formation.

The stability of a cartel is enlarged by a reasonably high income elasticity of demand. The durability of the cartei depends also on the long-term price (in)elasticities and income elasticity. In the long run substitutes can impede the operation of the cartel. The stability of a cartel is also ensured when the cartel members all have more or less symmetric development paths. Asymmetric development or dissimilar developments may, in time, hamper the cohesion of the cartel members.

The market structure attributes another important characteristic to cartel formation.[42] The first structural distinction is the number of suppliers and demanders. The main types of market structures that result from this classification are monopoly, oligopoly and perfect competition.[43] Mixed market forms between monopoly and oligopoly, respectively, and perfect competition add to the number of possible market structures.[44] Cartels are feasible in all market structures except for perfect competition and a single monopoly.

The second structural distinction is the degree of barriers to entry. Demand and supply can be characterized as open or closed and as approaching open or closed.[45] Cartelization is not necessarily restricted by conditions of open or closed supply (demand), depending on the purpose of the cartel. However, barriers to entry do make cartel formation easier to accomplish. The stability of a cartel can be endangered if market participants outside the cartel (in the same branch or from another branch) may easily enter the market.

The third structural distinction is the homogeneity or heterogeneity of the goods supplied

[40] Joint-ventures between market participants for research and development activities, limited knowledge sharing in government programs (like Esprit) are more probable than a cartel at this stage, although cartelization is possible.

[41] See section 2.

[42] Veröffentlichung der Preisbildungskommission des Eidgenössischen Volkswirtschaftsdepartements, Ergebnisse Eidgenössischer Untersuchungen, Bern 1957, in: H-H. Barnikel, *Theorie und Praxis der Kartelle*, 1972, p.214-225.

[43] A single monopolist can be defined as a firm or person who is completely alone on his/her side of the market, or who, if not alone, controls the market behaviour of the competitors.
A collective monopoly is an agreement among a group of competing price makers to restrain competition among themselves. Oligopoly is restricted competition among a small number of price makers whose competitive behaviour is restrained not by agreement but by their realization of their interdependence of their actions.
Perfect competition can be defined as a situation where all members of the market (buyers and sellers alike) regard the price as given and the necessary preconditions are large numbers of market participants and expertness on both sides of the market. T. Scitovsky, *Welfare and Competition*, Stanford University, Palo Alto, 1971, p.17 and Chapter 22.

[44] Veröffentlichung der Preisbildungskommission des Eidgenössischen Volkswirtschaftsdepartements, Ergebnisse Eidgenössischer Untersuchungen, Bern 1957, in: H-H Barnikel, op.cit., p.215/216.

[45] Open supply means either that a new market participant can enter the market at any time or that established market participants can increase their production capacity. Open demand can be described correspondingly. Closed supply prohibits both new entrants into the market and existing market participants from increasing production. 'Approaching' open or closed supply or demand describes the less extreme cases.

or demanded. A cartel can increase both the homogeneity and heterogeneity of supply, but cartel feasibility increases when homogeneous goods are involved.

The oil industry has witnessed several periods of cartelization. In 1928, Exxon, BP and Royal Dutch Shell established an international cartel when they concluded the 'As Is' Agreement and the 'Red Line' Agreement and instituted the 'Gulf Plus' price system.[46] The stability of the cartel was strengthened by the implementation in 1928 of a conservation policy in the US. This corporate cartel was a true accord among the main international oil companies, and the smaller companies behaved more or less in acquiescence. In 1947, this cartel was transformed into a new cartel and four more companies were included in it. The cartel operated without official acquiescence and was based on their joint operation of the concessions in the Middle East. According to Moran[47], the major achievement of the large international oil companies with regard to the oligopolistic control regime is: "(...) to use moments of harmony to set up structures that, once in place, would <u>automatically</u> give preponderance to those with greatest stake in long-term equilibrium, restrain the pursuit of short-term advantage, and penalize aggressive behavior without the need for direct retaliation. The structures were non-democratic, and created an apparatus of mutual self-denial that contradicted the idea of maintaining autonomy over individual decision-making. When the supra-sovereign mechanisms of restraint were in place, the oligopoly enjoyed remarkable success, despite threats, shocks, and challenges to its integrity. When a part of them began to disintegrate in the late 1950s, the oligopoly faltered." The downfall of the corporate cartel was, according to Johany[48], caused by the additions to total company and world reserves on the one hand, and the realization that their major concessions would expire on the other. Basically, the companies had a choice between increasing output, and thus reducing the prices, or maintaining high oil prices by means of output restrictions. Due to the imminent expiry of their concessions, a large volume of oil would certainly have been left unproduced. This in turn, influenced their user costs, discount rates and consequently their pricing goals. The individual oil companies did not have much choice in determining the desired output

[46] The names of the companies have changed through time. The names referred to here will be the current ones. The names of the companies have changed as follows: <u>Exxon</u> was first known as Standard Oil of New Jersey, then as Esso (in Europe they still use this name) and currently as the Exxon Corporation.
<u>Chevron</u> was first known as Standard Oil of California, then as SOCAL and is now known under its brand name Chevron. <u>Mobil</u> originates from two companies: Standard Oil of New York (SOCONY) and Vacuum. After they merged they were first known as SOCONY-Vacuum, then SOCONY-Mobil. <u>British Petroleum</u> was at first known as Anglo-Iranian, then as Anglo-Persian and currently as BP. In 1906 Koninklijke Olie (Royal Dutch) and the Shell Transport and Trading Company merged into <u>Royal Dutch Shell</u> (40% Shell and 60% Koninklijke Olie). <u>Gulf</u> was acquired by Chevron in 1984. <u>Texaco</u> never changed its name. Texaco acquired Getty Oil under disputed circumstances. Pennzoil subsequently filed a complaint, and the courts ordered Texaco to pay $3 billion for damages only. In 1989, the companies reached a settlement. The financial setback induced Texaco to sell some of its subsidiaries. Among the buyers was Saudi Arabia, who acquired some downstream activities and access to Texaco's distribution network on the US market (*Financial Times* 22-2-1988, p.22).

[47] T.H. Moran, Managing an oligopoly of would-be sovereigns: the dynamics of joint control and self-control in the international oil industry past, present, and future, in: *International Organization*, Autumn 1987, volume 41, no. 4, p.601.

[48] A.D. Johany, *The Myth of the OPEC Cartel, the role of Saudi Arabia*, University of petroleum and minerals, Dharan, John Wiley & Sons, Chichester, 1982, p.97.

level. The fact that the oil market gradually became more competitive meant that "any individual producer's efforts to prevent a price fall by output cutbacks would have only resulted in the reduction of his own revenues."[49]

With regard to the classification of OPEC as a cartel, some problems of definition remain unanswered. The existing literature is in this respect not particularly coherent. The cartel definitions cited in Cox[50] and Barnikel[51] do in general not include national states or international organizations as possible independent market participants. Several definitions are predominantly suited to cartels on national markets and do not capture the particulars of international cartels. OPEC would, according to quite a few of these cited definitions, not be a cartel.[52]

The dominant school of thought does, however, consider OPEC a cartel, although many times the terms oligopoly and cartel are used interchangeably. In contrast, the dissenting school of thought rejects the classification of OPEC as a cartel for several reasons. Mainly, economists from OPEC countries have tried to demonstrate that OPEC lacks several of the essential prerequisites to be considered a cartel.[53] Alnasrawi[54] for example states that: "The problem with a cartel theory as applied to OPEC is that it is imposed on the pricing behavior of an intergovernmental organization that lacks the necessary conditions and attributes of a commodity cartel. Economic theory on the subject of commodity cartels is explicit. In order for a cartel to exist in a market, certain features are called for, including an output quota system that determines the share of each member in the total output of the industry; an agreed-upon common price or price structure; and an agreement on quantitative and/or geographic market sharing. (...) More important, the governments of OPEC- member countries have always placed economic and political national considerations above the common economic interests of the group. It is these national interests that, in the final analysis, render fruitless any attempt to categorize an organization like OPEC. The major flaw in attempting to label OPEC stems from trying to impose the behavior of a firm-- which has profit maximization as its single most important goal-- on the behavior of a group of political entities, each one of which has different and numerous political, social and economic

[49] A.D. Johany, op.cit., p.97.

[50] H. Cox, op.cit., p.229-231.

[51] H-H. Barnikel, op.cit., 1972.

[52] Daoudi also surveys the various conflicting definitions which would or would not include OPEC within the definition. M.S. Daoudi, *The politactics of international cartels: economic illusions, political realities, and OPEC*, University Microfilms International, Ann Arbor, Michigan, 1985.

[53] Ibid., Chapter six; identifying OPEC, p.93-109. The author introduces a new definition to classify OPEC 'properly': "an organization with the potential and aspiration to become a cartel, which may behave at times as a cartel or may exhibit some of the characteristics of one, but which is not yet a full-fledged cartel. It is characterized by a loose framework of voluntary cooperation among its members, with great institutional flexibility in which members reach a broad general agreement on a general price structure, but without enforcing, formal production controls."

[54] A. Alnasrawi, *OPEC in a changing World Economy*, Johns Hopkins University Press, Baltimore and London, 1985, p.85 and 88.

objectives. Furthermore, each OPEC-member country has a different set of domestic and international priorities." Alsnasrawi's argument is not convincing with regard to the prerequisites of a cartel. A condition for a cartel is not based on motives, but on market behaviour of the market participants. His view that market behaviour of companies differs from market behaviour of a group of political entities, in itself a logical argument, is justified by motives of the market participants and not the market behaviour itself. Moreover, OPEC did operate a output quota system in the 1980s, and it has had a common agreed upon price structure and level.

The view that OPEC is not a cartel is also supported by Johany. He observes: "It seems to me (and one can only speculate in this regard) that the many able economists who are convinced that OPEC is a cartel just failed to recognize that the utility functions and constraints of the oil companies on the one hand and the host countries on the other hand are drastically different."[55] He consequently explains the price rise of 1973/1974 not in terms of cartel market behaviour, but as the result of the transfer in de facto ownership rights of oil deposits. The OPEC countries simply had lower discount rates than the international oil companies' effective account rate and therefore reduced output. This in turn increased the oil price, and consequently boosted (in the short-term) the oil producing countries income from exports. The increased income reduced the discount rates even more, especially since the marginal rate of return on domestic investment in the Gulf countries is much lower than the average world market rate of interest.[56] Johany's arguments, however, similar to those of Alnasrawi, are based on the discrepancy in motives of the international oil companies and the OPEC member states. He ignores the fact that the member states' market behaviour was concerted and did influence market conditions.

Peirce[57] also identifies problems with regard to cartel theory, OPEC and the usage of terminology. He observes that: "As OPEC achieved its initial successes in the early 1970s, economists turned to their models of the behavior of cartels, and other commentators tried to supplement such economic models with political or other explanations. (...) When OPEC began to have an apparent impact during the early 1970s, it was natural for economists to explain the phenomenon with the aid of the simplest model of pricing under competition and monopoly."[58] His analysis of cartels is based on the basic economic postulate that each individual tries to maximise his own utility. This postulate refers to the goals and the behaviour of individuals, not firms, cartels, nations or groups.[59] He then continues his argument that individuals will pursue their own interests within a group or bureaucracy or

[55] A.D. Johany, op.cit., p.97.

[56] Ibid., p.44-45.

[57] W.S. Peirce, *The Economics of the Energy Industries*, Wadsworth Publishing Company, Belmont California, 1986.

[58] Ibid., p.144.

[59] Ibid., p.144.

throughout political negotiations, and if the goals of the individual decision-makers are not identical with the objectives of the organisation, the "actions of the organization are likely to be inconsistent, reflecting the shifting alliances and compromises of the individuals involved."[60] With regard to OPEC he states that: "The fundamental problem with using the simple model of monopoly is that OPEC is a negotiating committee of countries in which about half of the world's oil was produced, not a unitary decision-making monopolist that can easily choose its rate of output."[61] According to Peirce OPEC differed from a standard manufacturing cartel because of its natural resource base, the limited absorption capacity of some member states' economies, and the exogenous cuts in production. These factors also enabled OPEC to maintain the stability of the cartel. The stability of the OPEC in the 1970s was, according to Peirce, further aided by "the incredible stupidity of the United States policy."[62] However, Peirce's assessment of the stability of OPEC was challanged in the 1980s. The factors that, according to him, strengthened the cartel in the 1970s, constituted a major strain on cohesion and stability of OPEC in the 1980s, when market conditions changed.[63]

Another approach to strategic market behaviour of (groups of) players is game theory. In game theoretic terms, the member states' different economic determinants characterise OPEC as a typical example of the prisoner's dilemma, i.e. a noncooperative game with conflict. This model assumes that the member states maximize their own utility functions, and, therefore, cannot make binding commitments. Although the member states can agree to cooperate, they will, when the time comes, act according to their maximum individual payoffs.[64] This opportunistic behaviour was also observed by Moran, who pointed out that the reward for cheating in OPEC was great.[65] The developments in OPEC in the 1980s confirm this view. This game theory model emphasises the instability of OPEC, which prevents them from coordinating their market behaviour as a single monopolist.

In game theoretic terms, one could say that in 1973, there were sufficient shared interests amongst the member states to agree, and thus they were able to reach a Nash equilibrium, because, given the market conditions in 1973, they were able to gain short-term supranormal profits and control of production. In this sense, 1973 must be considered as a unique moment, which could not be repeated. Moreover, 1973 immediately brought the asymmetry

[60] Ibid., p.144.

[61] Ibid., p.145.

[62] Ibid., p.146. On US policy he remarks on page 151: "This is not to imply that the policy was made by stupid people, but only that the net result of the compromises and political opportunism was a policy that did more to advance the interests of OPEC than of the consuming nations. The main themes of energy policy during the 1970s were (1) to prevent the substitution of other fuels for oil, (2) to keep prices low to consumers so they would not conserve oil, (3) to inhibit domestic production and refining, and (4) to prevent domestic producers of oil and gas from making large profits."

[63] See Chapter Five.

[64] E. Rasmusen, op.cit, p.29.

[65] *Supra* note 8.

among the member states to the fore, when they had to formulate their policy. This change was the result of the transfer of property rights, and, therefore, a change in the economic determinants of the member states' strategic behaviour and their ability to make credible commitments.

Game theory, however, does not sufficiently capture the political dimension of OPEC. OPEC is primarily an organisation of developing countries, and many times, also an organisation where the complex Middle East politics dominate. There have been situations after 1973 where agreement on production quota was realised, despite the fact that it was against certain member states' economic interests. This was mainly achieved as a result of political manoeuvring.[66] Thus, despite the existence of economic rewards for cheating for individual countries, the agreement was, at least for the duration of the political accord, credible. Member states with little interest in the political manoeuvring were compensated for their cooperation in economic terms. The member states with a stake in the political positioning provided this compensation, i.e. they transfered production from one country to another.

The lack of symmetry, different objectives and strategies does not entirely explain the instability of OPEC. Even in a asymmetry model of a leadership-follower type relationship between member states, the stability of the cartel would depend on the dominant player(s) establishing the leading price for the group, assuming there is a finite set of players. Jacquemin shows that in such a situation, in which the players take the interaction between their decisions and their environmental conditions into account, an equilibrium of the leader-follower type can be stable, without the tendency for the leader to disappear.[67] However, in OPEC, none of the players or sub-groups of players held after 1973 a dominant set of economic, political, and strategic factors, which facilitated a leadership role for a sustainable period of time, and gave the group some stability. For instance, economically Saudi Arabia and the other Gulf member states should have been able to assert a leadership role after 1973, but politically they lacked the power to do so. Consequently, in 1979/1980, Saudi Arabia was unable to assert its leadership, and prices were set by a group of players that lacked the economic parameters to assume leadership.

The debate over whether OPEC is or is not a cartel is surrounded by a certain degree of ambiguity. Most market behaviour studies are centred on the oligopolistic market behaviour of OPEC, and the term cartel does seem to be used with a certain casualness, a casualness that hints in some cases more towards an ideological choice than to fundamental theoretical logic. For example, Adelman[68] writes: "The organization OPEC is no more than a convenient *forum* for the constituent nations. They have formed a *loosely cooperating oligopoly* - or *cartel*, if you prefer two syllables to twelve. But no two cartels are the same.

[66] See Chapter five.

[67] A. Jacquemin, op.cit., p.72-74.

[68] M.A. Adelman, in: J.M. Griffin and D.J. Teece, op.cit., p.37.

They are all historical individuals, who change over time, and the *OPEC cartel* has changed greatly" (emphasis supplied).

The difficulty in analyzing OPEC in terms of cartel theory emanates from the fact that the governments of these nation states individually exercise public and political intervention in the market, which effectively limits the ability of OPEC to behave as a single monopolist. Moreover, the member states are endowed with different natural resources and factors of production, and their oil policy is part and parcel of the state's wider economic, political and strategic interests. This has consequences for the stability of the group, but also frustrates theoretical analysis of the group's market behaviour.

The problem with defining OPEC as a cartel is the fact that it is an intergovernmental organisation of nation states. However, in compliance with the cartel definition earlier supplied, national states or intergovernmental organisations can be considered as 'independent market participants that operate in the same or parallel sector, which are (potentially) competitors in the same market(s) or good(s), have action parameters in common and attempt to regulate or influence the market conditions to their own benefit.' An international organisation, through its statutes, could be seen as an agreement or accord between the independent market participants. The source of ambiguity is thus not the determination of certain market behaviour as cartel behaviour, but this behaviour cannot be explained by economic factors only. At this point, distinctions between market behaviour of nation states and/or private and public companies are vital. A cartel of nation states must then imply that, in terms of the earlier supplied definition, their 'attempt to regulate or influence the market conditions to their own benefit' should be read as 'to their own ***political, economic or strategic*** benefit'. This distinction might facilitate explanations with regard to the feasibility, the operation, and the durability of a cartel of nation states as opposed to cartels by companies. Nations states can apply a variety of measures to satisfy their own objectives, or the cartels' objectives. Thus, these measures can be used to reinforce the cartel but can also destabilise it. These measures can be political, economical, and ultimately also military. Based on its market behaviour, OPEC does comply, at particular moments in time, with the definition supplied, but the feasibility, operation, and durability of the cartel must and can be explained by more than economic factors only.[69]

5. Upstream market conditions

Exploration is by definition confined to those areas where the geological structure indicates oil deposits. Exploration for oil can be a complicated matter technically and is increasingly

[69] See Chapter 5.

capital intensive as more off-shore deposits are recovered.[70] The allocation of capital and labour to exploration is a function of the expected costs, the size of the concession areas and future price of oil.

The cost of production varies per oil deposit, depending on the depth and pressure of the well and the volume and quality of oil that can be produced. Furthermore, the costs of exploration used to be high in relation to the cost of production.[71] Presently, costs in some oilfields are concentrated in the production phase, reversing the cost relationship of exploration and production.[72] Investment in offshore production and secondary and tertiary recovery methods has initiated this reversal. The high proportion of fixed costs to variable cost are not only characteristic to the upstream industry, but also to a part of the downstream industry. Together with the large economies of scale in production and refining, they generated high barriers to entry in a period of declining or low nominal oil prices. New entrants had to meet high capital requirements in order to reap the benefits of the economies of scale in exploration and refining. The high fixed costs upstream and the substantial economies of scale increased the value of stable downstream markets for the producers and induced the refiners to put a premium on a continuous flow of crude oil. Oil producers and refiners were for these reasons compelled to internalize these costs, and integrated either forwards, backwards or both. By means of vertical integration they could (almost) evade market transactions, and the companies expected to gain from this, as they could save on transaction costs and accrue the profits of the integrated stage of production.[73] Vertically integrated companies could circumvent uncertainties and expenses of bargaining for raw materials on intermediate markets, minimize or avert the risk of dependence on (potential) competitors for their raw material supplies and in case of scarce supply of crude oil, and gain a competitive edge over non-integrated competitors in the downstream industry.[74] Once integrated and in control of the market, the seven major oil companies regulated their output and pricing. They determined both the internal price and prices for third parties at a high

[70] The development of the North Sea oil production posed a technical challenge to the companies. Previous off-shore oil production had not involved the geological and climatic circumstances and water depths as in the North Sea area. Exploration started in 1965, but was unsuccessful until December 1969 when the Montrose and Ekofisk fields were discovered. In the period 1965-1969, 21 dry holes had been drilled in the UK sector alone at a cost exceeding $50,000 a day. Development costs varied from $100 million for the Argyll field to $5,200 million for the Brent field. D. Fisher, *North Sea Oil; an Environment interface*, Universitetsforlaget, Bergen/Oslo/Tromso, 1981, p.32-33.

[71] "The governing factor in the economics of crude oil production is that exploration and drilling are expensive, while the actual cost of lifting the oil from the subsoil is relatively low. In other words, *capital investment is necessarily heavy, whereas current expenses are light*. This proportion of fixed and variable cost provides a set of circumstances characteristic not only of the production side of the industry, but also of some of its later phases." This citation is from P. Frankel, Essentials of Petroleum: A Key to Oil Economics, Chapman & Hall, London 1946; 2nd edition: Frank Cass & Co. Ltd., London, 1969, reprinted in: I. Skeet (ed.), *Paul Frankel; Common Carrier of Common Sense, a selection of his writings 1946-1988*, Oxford Institute for Energy Studies/Oxford University Press, 1989, p.3.

[72] Ibid., p.1.

[73] H.W. de Jong, Dynamische concentratietheorie, p.149.

[74] B. Levy, World oil marketing in transition, in: *International Organization*, Winter 1982, volume 36, no. 1, p.115.

level, and left only slim profit margins on refining.[75] This pricing policy benefitted the seven companies in the short-run because it enabled them to take their profits mainly at the upstream stage of production[76] and also deterred new entrants from entering the refining industry, or forced them to invest in both oil production and refining. Another incentive for the companies' pricing policy was the possibility to accrue large US tax advantages in the upstream phase. However, the higher oil prices in the 1950s attracted new companies into international oil production. Although the capital requirements were high and risky, the benefits were equally large in case of a successful find.

In the upstream stage the barriers to entry can be large because the acquisition of concession areas depends both on the availability and size of geologically promising areas.[77] The size of concessions can either limit or expand exploration expenditure and can deter new entrants from entering the market. The size of the concession is usually determined by private property rights or by governments.

In the US, the subsoil property rights are defined by the 'Rule of Capture', giving the owner of the land ownership of the subsoil natural resources also. The size of the concessions was until the 1930s small in relation to the cost of exploration and production. These small concession areas prevented investors from securing economies of scale[78] and stimulated excessive[79] production. The US market was as a result competitive until the large companies internalized these costs by extending their concessions over several tracts of land, and the government established rules with regard to production levels per oil deposit.[80]

In the 1920s and 1930s, some international oil companies managed to obtain very large concessions in the Middle East. A company would negotiate for sole exploration and production rights in a country and then decide whether to develop a deposit in joint-venture with other companies. In the case of Iraq, the British and French governments determined that the exploration and production rights should fall to companies from their own countries. The Iraq Petroleum Company was a joint-venture between these European oil companies. Diplomatic pressure from the US government on the British government ensured the participation of some American oil companies. Once the large companies had obtained and developed them, the enormous size of the concessions in the Middle East created substantial

[75] Oil price and production decisions, however, have to take competition from other fuels and final products into account. In some fuel and product markets there are no genuine substitutes for oil-based products, while in other markets oil products meet competition. This competition can also be government-induced, like in the case of coal and nuclear power.

[76] B. Levy, op.cit., p.115. It should be noted however, that the major oil companies only deterred new entrants in the upstream and downstream industry until the mid 1950s. (From 1960 onward, posted oil prices remained stable and declined in nominal and real terms.) During the 1930s, the cartel failed to deter Chevron and Texaco from entering the Middle East.

[77] The availability of attractive areas is dependent on the development of geological techniques, the size of the concession areas and the condition of the concessions.

[78] The Rule of Capture did, for this reason, stimulate firms to integrate horizontally and vertically.

[79] Excessive in the sense: "more than economically and technically viable".

[80] See Chapter 2.

barriers to entry.

The size of the concessions combined with the size of the oil deposits' production capacity, as in Saudi Arabia, might go beyond the companies' ability to process and market the oil independently, inducing them to produce and market the oil jointly or to market it through another companies' distribution channels.

In Saudi Arabia, for instance, 97% of the proven oil reserves in 1976 (180,862 million barrels) were within the Aramco[81] concession area, which was made up of 189,000 sq.km onshore and 31,000 sq.km offshore[82] Saudi Arabia proved to have the largest single oil deposits in the world[83] and the costs of production were very low.[84] Chevron obtained the concession in 1933. Due to delayed success in exploration and increasing spending, a joint-venture was created with Texaco. In 1948 both Exxon and Mobil joined Aramco because the magnitude of the production capacity was beyond the combined processing and distribution abilities of Texaco and Chevron. However, when Saudi output was increasing it could influence both price and market shares of other companies. The inclusion of Mobil and Exxon into Aramco in 1948 (drawing Chevron and Texaco into the framework of the international petroleum cartel) extended the potential to process and market this oil. This, in turn, also stimulated collusive market behaviour of the major oil companies.

In the 1950s and 1960s, the governments of new oil producing areas reduced the size of concessions and limited the number of contracts per company. This worked as a catalyst for increased exploration expenditure and lowered the barriers to entry in this particular phase. The North African countries wanted to limit the control of the large international oil companies in charge of the consortia operating in Iraq, Kuwait and Saudi Arabia. The governments, especially the Libyan government, not only decided to limit the size of the concessions but also to allot the concessions to a larger number of companies. This policy coincided with the desire of the smaller American 'Independents' and European state oil companies to expand in international oil exploration and production. These companies were attracted into international exploration and production because of the relative high (American) oil prices in the 1950s, the tax benefits in the US as a result of the 'golden gimmick', the low production costs of international oil wells compared to other oil deposits, and import controls, first voluntary and later compulsory, imposed by the American government, which subsequently cut off American refiners from cheaper oil.[85] The combination of new concession policies and the international expansion of smaller oil companies weakened the

[81] Arabian American Oil Company, a former joint-venture of Socal, Texaco, Mobil and Exxon.

[82] Saudi Arabian Oil & Gas Directory 1977-1978, p.162 and 168, in: *L'Arabie Saoudite d'hier à demain*, Cahiers 8-9-10-11-12, Centre de Recherches sur le Monde Arabe Contemporain, Universite Catholique de Louvain.

[83] Ibid., 162. The Ghawar oil field (241 km long and 35 km wide) is the world's biggest oil field (in 1977-1978), with reserves of about 68,000 million barrels.

[84] Saudi Arabia's crude oil reserves were in 1990 25.5% of total world crude oil reserves and about 65% of reserves of the countries around the Persian/Arabian Gulf.

[85] See Chapter 2.

market control of the major oil companies and kindled price competition. Although posted prices remained stabilised at $1.80 throughout the 1960s for tax purposes, reductions on the posted price increased to $0.30. In 1970 Libya's new leader Ghaddafi managed to expose the diminished market control of the major companies in the industry by pressuring the smaller oil companies to raise oil prices.[86] This led eventually in 1971 to the Agreements of Teheran and Tripoli, the first major success of the governments of oil producing countries united in OPEC, and the definite onset of a period of disintegration in the sector.

The decision to explore for oil and in which country, is a strategic decision for a company (or government). If successful, it will add to the crude oil reserves of a company (or country). Crucial for the exploration phase is to identify geologically attractive structures. Test wells must be drilled. If oil deposits are found, the commercial viability of the well must be established. The commercial viability of the well is based on the quality and quantity of the recoverable oil in relation to the exploration and exploitation costs and the expected price.[87] Crude oil differs in quality. This is expressed in the crude's gravity and the impurities it contains. The gravity of a crude is generally measured according to the American Petroleum Institute's (API) grading system. Each crude is given an API degree rating in relation to a specific gravity that has been given to water. Water has been rated as 10^0 API. Crudes lighter than water are given a higher rating. Crudes with a rating higher than 30^0 API are known as light crudes. Crudes with ratings between 22^0 and 30^0 API are considered as medium crudes. Crudes below 22^0 are heavy crudes.

The gravity of a crude is important in terms of processing possibilities and in terms of costs of processing. Refining of a light crude yields more products at the lighter end of the barrel. Depending on the market for oil products, the refinery needs to balance the input of crude with the output of oil products. The impurities of crudes are also an important factor in the attractiveness (and price) of certain types of crude. The content of sulphur, which can vary from 0.1% (called a sweet crude) to 7% (called a sour crude), and/or metals incurs extra costs to processing. Environmental legislation to constrain the content of sulphur in gasolines has made sour crudes less attractive (more costly in existing plants) to refine.

Supply of sweet light crudes has become relatively scarce. New additions to reserves from the 1970s onwards were mainly additions to heavy crude reserves, bringing the share of heavy crude to approximately 85% of total reserves.[88] The quality of oil has become more and more important as demand for oil products shifted to the lighter end of the barrel and refinery technology advanced. Demand for a particular crude oil and its expected price is therefore dependent on the demand for oil products on particular markets and the technical

[86] A. Hammer with N. Lyndon, *Hammer, witness to history*, Coronet Books, Hodder and Stoughton, Sevenoaks, 1987, p.443-479.

[87] The commercial viability of a well is not only based on the actual costs of exploiting the well, but also on political risks. In general, the evaluation of the political risks will precede the decision to take an option on the concession and the test well drilling.

[88] *OPEC Bulletin*, March 1982, p.1.

possibilities of the refining industry to refine the various qualities of crudes. Secondary conversion upgrades the yield of lighter products from heavier crudes. Secondary conversion does, however, involve costs.

The impurities and differences in grades make crude a not completely homogeneous good. With the use of a single and later, a dual basing point system, a 'marker crude' and price differentials between types of crude, the industry has tried to homogenize crude oil. The pricing systems proved to be unsustainable because they always served the compatibility of some types of crude better than others. The pricing system of the international petroleum cartel advanced American crudes, the post-war posted prices stimulated consumption of the Middle East light crudes and the OPEC pricing system is still in limbo with regard to price differentials.[89] Changes in price differentials had a substantive impact on the competitiveness of one crude over another and impeded market shares of certain crudes. OPEC has been unable to deal with this problem efficiently.

6. **Downstream market conditions.**

Refining is in many ways the lynch-pin of the oil industry. In a refinery, crude oil is processed into semi-finished and finished products. The exact proportions of the fractions (light, middle or residues) of a processed barrel of crude depend on the type of crude and the configuration of a refinery. The configuration of a refinery is also influenced by the specific demands of a market. It is characteristic of this stage of production that the fixed cost is high in relation to the variable cost.

Refining is not geographically confined to certain areas or countries. This sub-sector has the flexibility to produce where factor prices are advantageous and at the lowest possible production costs. As a result three kinds of refineries can be distinguished: 1) source refineries; 2) market refineries; and 3) export or intermediate refineries. A source refinery is based near the oil wells, and the oil products are either locally consumed or exported to other consumer markets. The market refinery processes the crude oil into products near its main consumer markets, and the export refinery is neither based near the oil production centres nor near the consumer markets. The location of an export refinery is usually located in a coastal area at major (sea)trading routes or distribution centres, like Singapore and formerly Aruba and Curacao.[90] The location of refineries is dependent on several factors. These factors can be location specific, ownership specific and/or internalization-incentive

[89] See Chapter 5.

[90] The export refineries are currently re-located to the coastal areas in the oil producing countries. The advantage of the traditional export refinery centres, which served as major distribution centres, has evaporated after 1973. The structural over-capacity in world refining in the 1970s and early 1980s, and the subsequent restructuring process, has resulted in the closure of these refineries by the international oil companies. In some cases, the host governments decided to continue operation of the plants to maintain employment. However, the outlook of these plants was dubious.

advantages.[91]

Location specific advantages can be classified as: spacial distribution of natural and created resource endowments and markets; prices, quality and productivity of inputs, i.e. labour, energy, materials, semi-finished products; international communication and transport costs; investment incentives and disincentives; trade barriers; infrastructural provisions; economic system and government policy; and the institutional framework for resource allocation.

Ownership advantages can be accrued from barriers to entry, i.e. property rights, innovations, human capital of the firm, organizational and marketing systems, and the advantages that arise from the coordination of separate, but complementary activities of the firm at low or zero marginal costs, i.e. information, organization, access to inputs, capital, risk diversification.

Internalization-incentive advantages are generated by market imperfections. The incentive to internalise depends on the relationship of four sets of factors: 1) industry-specific factors; 2) region-specific factors; 3) nation-specific factors; and 4) firm-specific factors.[92] They include among others: capture of economies of interdependent activities; the avoidance or exploitation of government intervention; security of supply; control of the conditions of sale of inputs; to control distribution; and to engage in transfer-pricing.

The location of a refinery can also be influenced by the price of crude oil fob (free-on-board) and cif (cost, insurance, and freight) and the price of semi-finished and finished products in relation to transportation costs. Transport has been an influential sector in the petroleum industry.[93] The major markets for oil and oil products are, except for the American market, distant from the main producing areas. Several types of transportation can be distinguished depending on the location of the crude production areas and the decision where to process the oil. There are two main types of transport: transport of crude oil to refineries (source, export/intermediate and market refineries) and transport from refineries to the consumers. The transport of crude oil and oil products takes place by pipelines, ships, railways and road transport.

Pipelines are a means of transport in producing as well as consuming countries. Pipelines are constructed to transport crude oil from oil wells to source refineries, from wells to harbours, from harbours to market refineries and from refineries to distribution centres and (large) consumers. The seaborne trade is predominantly crude oil transport, although the share of oil products is increasing. Rail and road transport is used for transporting oil products to the (small) consumers.

The importance of these different types of transport has changed over time as the location

[91] J.H. Dunning, *Explaining International Production*, Unwin Hyman, London, 1988, p.19-39.

[92] N. Hood and S. Young, op.cit., 1979, p.56.

[93] Standard Oil Trusts' market control was based on the control of the transport of crude oil and oil products in the United States. Samuel's Shell company was also a trading company. After Shell merged with Royal Dutch, the strong position in the transport sector gave the company a competitive edge over the other companies.

of the refining industry has altered. Up to 1945 refineries were located in the oil producing areas. These source refineries exported their products to the oil product markets in consuming countries. The product markets (outside the US) were very small at that time. The main fuel in Western Europe (and Japan) was coal. The market size for gasoline, for instance, allowed only a small refinery capacity and there was no demand for the full range of refinery fractions on the European market. Transportation of separate oil products from source refineries to their main export markets was more efficient, since no waste products had to be transported.[94] The importance of source refineries manifested itself in the supply pattern of oil products before 1945. Until the First World War, oil products were exported from the US to meet a growing European demand. After 1930 a growing volume of oil products from the Caribbean and Arabian/Persian Gulf refineries was exported to Western Europe.[95]

After 1945 the European governments stimulated the build-up of the oil processing industry in their countries to meet the increasing needs for energy. When the supply of coal fell short of demand, the European countries had to import increasing volumes of oil products.[96] These imports depleted their foreign exchange reserves and influenced the balance of payments negatively.[97] The governments of some Western European countries, notably France, Italy and the Netherlands, implemented a policy to improve the incentives to investment in refineries in their own countries. The new market refineries facilitated the growing demand for oil products. The governments imposed tariffs on imports of oil products, once they had reached a reasonable rate of self-sufficiency.[98] The oil companies were, therefore, more or less compelled to invest in the oil processing industry in these countries.[99] However, the increased use of oil as a main fuel in Europe and the market opportunities for a variety of oil products made it more efficient to process oil close to consumer markets in any case.[100] The competitiveness of the market refineries in Western Europe was enhanced by the transport cost advantages of crude oil transport (as bearer of

[94] W. Molle and E. Wever, *Oil Refineries and Petrochemical Industries in Western Europe; Buoyant Past, Uncertain Future*, Aldershot, 1984, p.29.

[95] Ibid., p.29.

[96] Studie-comité van de Westeuropese Steenkolenproducenten, *Kolen en Energiepolitiek*, Brussel, 1955, p.13.

[97] The European currencies were not convertible until the late 1950s. The states had to deplete their dollar reserves to pay for their increasing energy needs which could not be met by the local coal production.

[98] France, Italy and the Netherlands reached the point of self-sufficiency in oil products already in 1950. Great Britain reached this point only in 1980; Belgium in 1970; Greece in 1975 and Spain in 1970. Portugal, Luxembourg, Ireland, Denmark and West Germany never reached self-sufficiency for refined oil products. W. Molle and E. Wever, op.cit., p.45.

[99] R. Lefeber and J.G. van der Linde, Europese Integratie vergt een energie(k) beleid, in: *SEW* (6) juni 1987, p.452/453. The alleged political risks, especially in the Middle East, also made the oil companies eager to process the crude oil close to the consumer markets. They avoided risking their capital intensive production processes to nationalizations.

[100] "The consumption of oil products in Western Europe increased from approximately 60 Mtons in 1950 to some 619 Mtons in 1975." W. Molle and E. Wever, op.cit., p.33.

several commercially attractive finished products) over oil product transport.[101] The international oil companies (and at a later date also smaller companies) duly changed their investment pattern in favour of the market refineries. In 1953/1954, the large international oil companies and the French company CFP owned approximately three quarters of Western European refinery capacity.[102] As well as enjoying (partly government induced) advantages of market refineries over source refineries, the international oil companies were also prompted to alter their downstream investment policies to avoid political investment risks. The growing nationalist movements in the oil producing developing countries, kindled by the de-colonisation process[103], incited the international companies to spread their investment risks and to capture the advantages of the changing European demand pattern at the same time. Oil trade and transport patterns therefore changed after 1945 from oil product trade and transportation to crude oil trade and transportation. In 1971, 6,555 billion tonne-miles of crude oil were traded seaborne compared to 900 billion tonne-miles of oil products. Oil (crude and products) constituted 64% of all seaborne trade (11,730 billion tonne-miles) in 1971.[104]

The refining industry entered a troublesome period after 1973 as profit margins narrowed due to structural overcapacity, and as the international oil companies lost control over oil pricing. The annual average growth rate of the refining industry was 8% before 1973. The buoyant growth had induced many companies to increase their refinery capacity.[105] After 1973 demand slackened for refinery products, while the refinery capacity was still increasing as a result of previous investment decisions. The considerable time-lag between the investment decision and the actual production of the new refinery caused extra capacity to come on stream after 1973/1974, by which time the market outlook had already changed substantially.

In the 1950s and 1960s, demand for fuel oil was strong and the refineries were equipped, given the then current chemistry of supply, to serve this market. From the 1970s onwards, demand shifted to the lighter end of the barrel, when demand for fuel oil by electricity generating plants decreased due to higher prices, and demand for gasoline increased. At the same time, the chemistry of supply changed in favour of heavier crudes, which necessitated secondary refining to supply the markets with a sufficient spread of products. These

[101] Ibid., p.30.

[102] United Nations, Secretariat Economic Commission for Europe, *The price of oil in Western Europe*, E/ECE/205, Geneva, March 1955, Annex (e) chart 3. Shell, Exxon and BP owned the large share (around 50%) of refinery capacity in Western Europe.

[103] In 1953 Mossadecq of Iran nationalized the oil production of British Petroleum and the Iranian National Oil Company took over. The intervention of the United States (CIA) and the refusal of the international oil companies to market any of the Iranian oil resulted in a denationalization and take-over by a consortium of international oil companies.

[104] S. Sinclair, *The World Petroleum Industry; The market for Petroleum and Petroleum products in the 1980s*, Euromonitor Publications Ltd, London, 1984, p.337.

[105] J. Boers, G. Eisenloeffel, C. van der Linde, Olieraffinage en petrochemie in de Golfstaten en EG, in: *ESB*, 22-2-1984, p.190-195.

fundamental changes in supply and demand affected the profitability of market refineries compared to source refineries, especially when the oil producing countries decided to refine lighter crudes themselves. The oil producing countries (especially OPEC countries) embarked on large investments to increase their refining capacities.[106] The economic and technical possibilities of separating primary and secondary conversion improved the refining industry's flexibility.

Although the number of independent refiners in the consuming countries had already increased in the 1960s, the bulk of refinery capacity is undertaken by vertically integrated oil companies.[107] Some smaller integrated companies are crude oil producers and refiners with mainly distribution outlets in the fuel markets. The large oil companies, however, are vertically integrated from crude oil production to petrochemical activities, including a world-wide refinery capacity with distribution networks in the fuel markets.[108] Companies like BP, Exxon and Shell are also important petrochemical manufacturers and compete with chemical companies on their markets.[109] The move upmarket of these large international oil companies materialized in the 1960s.[110] Their diversification efforts into wider energy sources (notably the production of coal, gas and in some instances also nuclear power) in the 1960s and 1970s was also significant.[111] Their initial limited success in diversification,

[106] In 1960 OPEC's share in refinery capacity was 2.2 million barrels per day, in 1970 3.6 million barrels per day, in 1980 5.8 million barrels per day and in 1986 6.6 million barrels per day. The OECD share in refining capacity was in 1960 15.9 million barrels per day, in 1970 32.8 million barrels per day, in 1980 47.4 million barrels per day and in 1986 36.5 million barrels per day. *OPEC Facts and Figures 1986*, Vienna, 1987, p.11.

[107] The participants in the subsequent stages of the oil industry are differentiated. Private as well as public companies participate in all phases of production of the international oil industry. The country of origin of the public and private oil companies can be either a producing country or a consuming country. Companies can be active in one phase of production, several phases of production or all phases of production. The latter companies are vertically integrated and the former are not. Companies with activities in one or two phases of production are either small or horizontally integrated companies. A few large companies are both horizontally and vertically integrated. The activities of these large international oil companies even extend to the exploitation of other sources of energy and to activities in the chemical and bio-technology industry, which are not strictly related to the oil industry. The corporate strategy of the various companies in the oil industry will differ according to their position in the industry.

[108] "(....) the share of petroleum companies in the production of basic petrochemicals in the United States is increasing. In 1974 the share of that category in total production was 65 per cent, in 1980 it had already increased to 70 per cent. The height of this share illustrates once more the close relation between oil refining and the production of basic petrochemicals. That the two largest producers in the world of ethylene and benzene are the largest oil companies (Shell and Exxon) is certainly no coincidence." W. Molle and E. Wever, op.cit., p.74.

[109] In Western Europe the share of oil companies in total ethylene and benzene capacity in 1980 was respectively 42% and 46%. W. Molle and E. Wever, op.cit., p.75.

[110] "The most recent expansion has been the move into oil-related areas of operation. In the 1950s and early 1960s the companies moved heavily into the petrochemical business. This was a natural step, since they owned the basic feedstocks and had large quantities of the necessary investment capital, and also since the chemical industry uses processing patterns much like those they were familiar with from petroleum refining. As a result, by 1962 the oil companies owned or operated over one-third of all U.S. petrochemical plants, while petrochemicals accounted for over 60 percent of the value of all organic chemicals produced in the United States." M. Tanzer, *The Race for Resources; continuing Struggles over Minerals and Fuels*, Heinemann, London, 1980, p.106.

[111] Ibid, p. 106-107. The oil companies began entering the competitive fuel industries in the early 1960s. "Their mixed success in earlier petrochemical ventures had reinforced the historic lesson that the companies' greatest advantage was to be gained from becoming monopolistic suppliers of raw materials; it is in such industries that one can get the highest monopoly profits, because control of raw materials can bar new entry into an industry,

coupled with the need after 1973 to restructure their core activities, interrupted their diversification strategies. In the 1980s, the large oil companies had, for the time being, retreated to their 'core' activities after some of their acquisitions outside the oil and gas, or even energy, sphere turned out to be loss generating ventures.

The diversification of their upstream and the upgrading of downstream activities became more important after 1973 when the oil producing countries increased the capacity of source refineries and developed plants to produce bulk petrochemical products like ethylene and methanol (important inputs for the petrochemical industry). The build-up of capacity in these countries invoked a restructuring process of the refining industry in the consuming countries.[112] Technological advances in refining and transport facilitated decentralized production. The refining industry can be divided into the processing of oil into bulk products and the processing of oil into advanced high value added oil products. These two refining processes do not necessarily have to be located in the same vicinity. The transportation costs of and demand for bulk oil products determine the best 'mix' for a refinery configuration.

The structural overcapacity in world primary refining capacity increased when more oil products were produced in the oil producing countries in the 1980s and the restructuring process of market refineries was not yet completed.

In 1978, world (excluding centrally planned economies; CPE) refining capacity utilization was 74.3%, in 1982 65.4% and in 1988 78.2%.[113] Capacity utilization in the CPE's was in 1978 88.3% and in 1988 82%.[114] Exxon, Shell and Mobil and some state oil companies from consuming countries (notably from Japan) increased their primary refinery capacity in joint-ventures with state (oil) companies of the oil producing countries (mainly in Saudi Arabia) and closed down market refineries and upgraded others in their main consumer markets.[115] In Western Europe, restructuring of the refining industry in the 1970s and 1980s resulted in decreased primary refining capacity (See Figure 1.1) and an emerging specialization in complex refining. In 1978, Western Europe still had the largest refining capacity in the world, but with the lowest utilization rate (63.1%).[116] Within 10 years, the refining capacity had diminished by 31.4% and the utilization rate had improved to 73.6%. In 1990, North America had the largest share in refinery capacity, followed by the Soviet Union and Eastern Europe, Western Europe and Asia/Australasia (See Table 1.2).

The recent restructuring of the world refining industry is not without an impact on the predominant type of transportation. The emphasis on crude oil shipping will decrease

thereby allowing very high rates of return on investment."

[112] J. Boers, G. Eisenloeffel, C. van der Linde, op.cit., p.190-195.

[113] Capacity utilization in world excluding CPE: 1978: 74.3%; 1980: 70.1%; 1982: 65.4%; 1984: 71.3%; 1986: 75.3%; 1988: 78.2%. Calculations based on *BPSR 1989*, p.17.

[114] Capacity utilization in the CPE's: 1978: 88.3%; 1980: 85.8%; 1982: 80.2%; 1984: 80.1%; 1986: 81.1%; 1988: 82.0%. Calculations based on *BPSR 1989*, p.17.

[115] J. Boers, G. Eisenloeffel, C. van der Linde, op.cit., p.193.

[116] Calculations based on *BPSR 1989*, p.17.

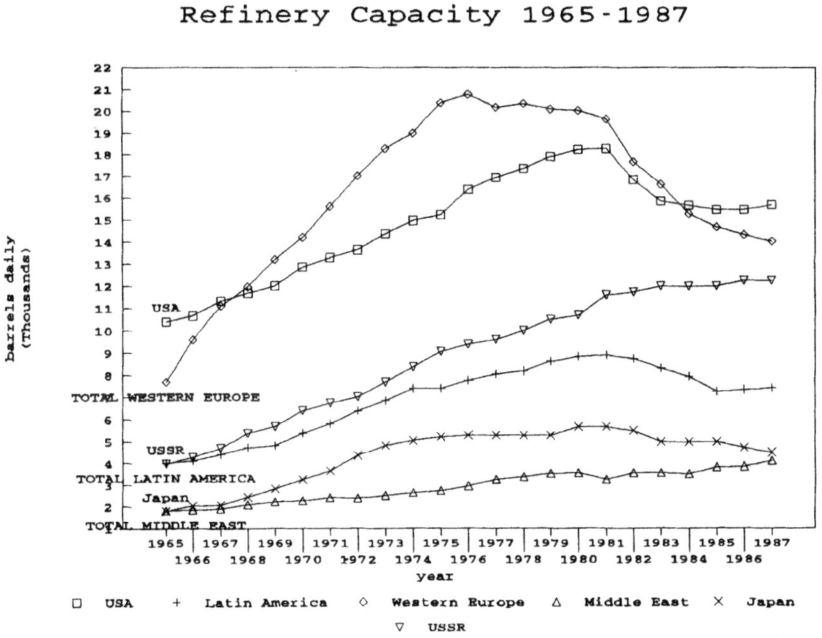

Figure 1.1

relatively, and transport of oil products (intermediate products mainly) will gain in importance. The transportation phase is thus a function of the location of the refineries, but the decision where to locate refineries is also related to transport costs and techniques.

Refining is typically a case of joint production, which makes it difficult to equate supply with demand exactly. The internalization of costs of the joint-production enhances vertical integration. Refining, given the input of a certain brand of crude oil and refinery configuration, produces a fixed volume of different finished and semi-finished products. The demand patterns of this variety of (semi) finished products, such as petrol, kerosene, fuel oil and naphtha, deviate. The balancing of refining output with differentiated markets for these products requires a wide distribution network. Supply is more or less fixed by the refinery configuration and types of available oil, while demand for oil products is determined by the conditions on the sub-markets.

The main market for fuel oil is electricity generation.[117] Up to 1973, fuel oil was

[117] Pricing of the individual petroleum products is complicated because the products are produced in a joint production process. Usually pricing of oil products is a reflection of the price of its near substitutes and not at all times the cost of refining of a particular product yield. In the 1950s gasoil was priced very low in order to stimulate electricity plants to use fuel oil instead of coal. In markets without a real substitute, like petrol, gasoil and kerosene prices could be set higher to compensate for lower prices on products with near substitutes.

Table 1.2 Refinery capacity in 1978, 1984 and 1990 in thousand barrels daily

	1978	1985	1990
North America	19,715	17,535	17,440
West Europe	20,355	15,200	14,190
USSR/East Europe	14,575*	14,435	15,275
Latin America	8,175	7,425	7,955
Middle East	3,375	3,810	5,050
Africa	1,730	2,555	2,885
Asia & Australasia	10,120	12,825	12,980
Total	78,045	73,785	75,775

*:including other CPE's; source: BPSR 1980, 1991.

increasingly used, in substitution for coal, to generate electricity.[118] After OPEC raised the oil prices in 1973/1974, coal regained its attraction for electricity generation. The supply of cheap coal from some large coal exporting countries increased steadily as a result of investments by governments, as well as some private companies (oil companies included).[119] Further, the redefinition of energy policy goals by oil importing countries, especially after 1979, stimulated substitution away from oil. In order to reduce the dependency on imported oil, especially oil from the OPEC countries, they obliged their (mostly) public electricity generating companies to decrease the use of oil to a certain prescribed level. This policy was especially successful in some Member States of the European Community.[120] France, Belgium, the United Kingdom and to a lesser extent the Federal Republic Germany increased their efforts to generate electricity in nuclear power plants.[121] Total world consumption (excluding CPE's) of fuel oil contracted (from 685.2 million tonnes in 1978 to 446.2 million tonnes in 1988[122]) and distorted the output balance in traditional configurated refineries compared to demand.

Between 1978/1979 and 1983, the markets for gasoline contracted in the US, Western

[118] R. Lefeber and J.G. van der Linde, op.cit., p.452-453.

[119] "Thanks to the careful acquisition strategy that began 15 years ago, 11 oil companies now own 25% of all coal in the country (i.e. United States, author) and the oil industry has the future of coal in its grip. And big oil has big plans for coal. From a modest 22% production share in 1978, oil companies have embarked on an ambitious expansion program that will put their share of national production at almost 50% by 1985." M. Tanzer, ibid., p.113. The share of US coal production in 1980 in world production is around 22%. J. W. Biekart, Steenkool en Macht, in: *Intermediair*, 1 February 1985. However, in the second part of the 1980s, the diversification strategy of oil companies changed and the companies have returned to a strategy to strengthen their core activities. Take-overs by oil companies are mainly concentrated on other oil companies. The acquisition of cheap oil reserves on Wall street has been a major incentive for their investment policy.

[120] R. Lefeber and J.G. van der Linde, op.cit., p.462-463.

[121] Ibid., p.463.

[122] *BPSR 1989*, p.12.

Europe and Japan. In most industrialized countries governments increased taxes to decelerate the rate of growth of gasoline consumption.[123] In the developing countries and the CPE, such policies were either absent or poorly implemented and gasoline consumption continued to increase.[124] World gasoline consumption decreased from 653.6 million tonnes in 1978 to 609.9 million tonnes in 1982 and increased from 1982 onwards to 688.5 million tonnes in 1988.[125]

The development of new oil products and new production processes induced companies to expand beyond the boundaries of their main sectoral activities.[126] Oil companies advanced progressively into the petrochemical industry. In the early 1970s, more than 90% of the basic organic chemicals were derived from oil or gas feedstocks. The growing importance of the petrochemical industry as a consumer of oil is reflected in its growing share in NCW crude demand. In 1960, the petrochemical industry consumed 2.7% of total NCW demand, compared to 7.5% in 1988. [127] This made the boundaries between chemicals and petrochemicals also indistinguishable.[128] The importance of petrochemicals for the oil companies is reflected in the share of income of chemicals to total income. In 1989, 25.3% of total income of Shell was achieved in the chemical sector. Exxon's income from chemicals was 30.8%, BP's 18.4%, Amoco's 29.9%, Mobil's 23.7%, Chevron's 29.5% and Texaco's 11.1%.[129] The refining and petrochemical industry are exemplary. The refineries and the petrochemical plants of the large international oil companies are usually operated in large industrial complexes.[130] The advantages were minimum transport costs and security of supply. Often these two industries are, because of the output/input linkage, part of a vertically integrated firm. The petrochemical industry processes refinery products and other hydrocarbons, like coal and gas, into a wide range of final products. Fertilizers, insecticides, paints, polymers, plastics, synthetic fibres etc. are products from the

[123] In Europe the government taxes constitute 65% of consumer gasoline prices (f.i. in the Federal Republic of Germany 59% and in France 72%), in the USA 32% and in Japan 43%. *OPEC Facts and Figures 1987*, p.26.

[124] S. Sinclair, op.cit., p.29.

[125] *BPSR 1989*, p.12.

[126] A similar development can be distinguished in other industries. The branch boundaries need continuous reassessment as a result of industrial product and processing developments. W. de Haan, C. van der Linde, F. van der Molen, Een zwaargewicht slankt af: Unilever 1965-1980, in: F. Crone en H. Overbeek (ed.), *Nederlands Kapitaal over de grenzen*, SUA, Amsterdam, 1981.

[127] H. de Ruiter, *Oil and Chemicals-do they mix?*, address to the National Association of Petroleum Industry Analysts/Petroleum Investor Relations Association conference, Tucson, 6 October 1989, Group Public Affairs, London, October 1989, p.12.

[128] R.B. Ballance, *International Industry and Business, Structural change, Industrial Policy and Industry Strategies*, Allen & Unwin, Boston/Sydney/Wellington, 1987, p.150.

[129] *Petroleum Economist*, July 1989, p.9.

[130] In the United States natural gas is used as the main fuel and feedstock in the petrochemical industry. In Western Europe and Japan naphtha is the main fuel and feedstock. In OPEC countries the oil processing industry is increasingly using associated gas. Until recently this gas was flared off as a waste product in the production of oil. This difference between the main petrochemical markets is based on the energy endowments of the countries.

petrochemical industry. Not only the number of finished products but also the utility of the products makes the petrochemical industry a very important producer of inputs for other industrial production processes. The petrochemical industry manufactures a growing share of refinery output, thus upgrading the use of oil.[131] The oil companies produce relatively more heavier petrochemicals and plastics than other chemical companies because of their access to refinery stream for feedstocks. However, this makes the chemical operations of the oil companies more dependent on oil price movements. Despite rationalisation of their petrochemical capacity and diversification efforts, the chemical companies are still less dependent on oil price movements. They are more diversified and have geared their operations in the faster-growing, high technology sectors of the industry.

In the oil exporting countries the petrochemical industry increasingly uses associated gas as a fuel and feedstock. As a by-product of oil production it was previously flared-off. The advantages of associated gas are obvious from the perspective of production costs. However, the oil production in most oil exporting countries, notably OPEC countries must remain at a certain level to secure the necessary feedstock for the petrochemical industry. This has created problems for the industry in Saudi Arabia where the production of the petrochemical industry came on stream in a period when oil production was severely restricted in order to stabilize crude oil prices.[132] The complex relations between raw material production, refining and petrochemical industry on the one hand, and retaining acceptable crude oil prices and income from crude oil exports on the other, has hampered the execution of OPEC policies since 1973 and added to the confusion on the oil market.

7. Conclusion

The oil industry developed from a lamp oil producing industry into a fuel and, more recently, also into a petrochemical products producing industry. The number of products derived from oil increased over time, and the production line of oil became longer and more complex. Crude oil developed increasingly into a noble fuel and feedstock.

The market conditions in the separate stages of production are counter-cyclical. This induces bottlenecks between the various stages of production, which necessitate continuous adjustments. Market participants have sought to control the resultant fluctuations by vertical and horizontal integration, collusive market behaviour, and cartel formation.

[131] Naphtha, kerosene and LPG are feedstocks for the petrochemical industry. "In the period 1950-1974, the industrial sector climbed from about 10 per cent to about 30 per cent of total use, an increase that was largely due to the petrochemical industry demanding more and more products as feedstocks for its production processes." W. Molle and E. Wever, op.cit., p.33.

[132] The level of Saudi oil production provides for a related level of associated gas. Severe cutbacks in oil production could endanger the raw material and energy supply of the petrochemical industry. The industry could lose its competitive advantage of cheap feedstocks because of Saudi production quota, arranged under the OPEC price stabilization formula. Additionally, government investments suffer from the decreased income from oil exports. The Saudi budget depended in 1985, just before the drop in oil prices in 1986, for 90% on oil income. C. van der Linde, Olie: de 'smeerolie' van het international marktsysteem, in: L.C. Biegel, *Zicht op het Midden-Oosten*, Boom/Venture Press, Amsterdam, 1985, p.177.

Upstream market conditions are determined by a set of technical and economic factors. The decision to develop a field depends on the geological prospects of an area, the size and the depth of the wells and the quality of oil. These factors determine, together with capital and labour costs, the production costs. Other factors that are taken into account before exploration starts are the infrastructural circumstances, and the proximity to processing centres and consumer markets. The investment decision is based on these estimated costs and their relation to future demand and prices. The assessment of future demand and prices also depends on downstream market conditions, inter-fuel competition and expected economic growth in the consumer markets. The investment decision is also influenced by government policies, on, for example, taxation, concessions or exploration and production contracts, ownership rights, trade policy, energy policy, and, of late, environmental policy.

The lynch-pin of the oil industry is in many ways the refining industry. Oil is processed into semi-finished and finished products in a refinery. The refinery should ideally match the chemistry of supply to the chemistry of demand. The composition of the refined barrel depends on the quality of the oil input and the technique of an installed refinery, suggesting output flexibility. However, once refining capacity is installed, the composition of the refined barrel becomes rather inflexible and can be a constraint to adapting effectively to changes in demand. The same inflexibility existed with regard to changes in the chemistry of supply.

In general, refineries built before 1973 were specialized in processing light crudes. Lighter crudes were cheaper to process and in ample supply. In the 1970s, the refineries were facing both a change in the chemistry of supply and in the chemistry of demand. The additions to the reserves were mainly additions to the heavy crude reserves, and the producing countries earmarked lighter crudes for processing in their own refineries. This put a premium on lighter crudes. In the 1970s, the production costs increased both as a result of the crude oil price increases and the increases in differentials between light and heavy crudes.

The shift in demand to the lighter end of the barrel, in part induced by the consuming governments' energy policies, further increased costs for the refiners. The rather fixed composition of a refined barrel can unbalance the product markets and increase price volatility. Meeting increased demand for one product can create a surplus in other product markets and depress prices in that market. When this increased demand is not met, price levels in that particular product market increase, and this may lead to an undesired suppression of demand. The structural changes in supply and demand made a major restructuring process in the refining sector necessary. The response to these changes was rather slow, and it took some time before the changes were acknowledged as structural. Some governments had, in the mean time, offered the refining industry some protection through subsidies, trade barriers, and price controls. Small independent refineries and some state-owned refineries were less able to raise the necessary investment capital and lobbied for continued protection. In the US especially, protection of the small refineries led to a delay in the restructuring process in the 1970s. The large international oil companies responded more quickly and rationalised their downstream industry to suit both the new market conditions and their own upstream operations.

The market developments and market structures in the different stages of production are clearly interrelated. Although this study mainly concentrates on the changes in the market conditions and structure of crude oil production and consumption and does not attempt to offer a full account of market developments in the separate product markets, the relationship between the upstream and downstream sectors is the focus of this study.

Chapter Two

Market Developments in the United States 1860-1940

1. Introduction

The industrialization process in the second half of the nineteenth century in Europe and the US was accompanied by the growth of mineral fuel consumption. Although wood still dominated energy consumption in 1880, mineral fuels were rapidly gaining importance. In the first instance coal developed as the major energy input in these economies.

In the US coal was, in the first half of the twentieth century, progressively substituted by oil, while in Europe coal remained the dominant fuel until after the Second World War. This deviation in the development of mineral fuel consumption between the two regions can be explained by the diverging energy endowments of the respective regions. The absence of commercially exploitable oil and gas resources in Europe, and the subsequent protection of the domestic coal mining industry, hampered the early expansion of oil and gas consumption in Europe. In other parts of the world, the development of regional and world demand, transportation, and in general the local economic infrastructure, determined the rate of development in the oil industry. These particular market conditions greatly influenced the pattern of the international oil market developments, and thus the course of the crude oil growth cycle.

The US was endowed with large, commercially attractive, coal, oil and, gas resources. From 1859 onwards, the US developed into the main producer, consumer and exporter of oil and oil products in world. Only after 1945 did consumption and production in other regions of the world expand at a faster rate than in the US. The US oil market matured much earlicr than other markets in the world.

US energy requirements for industry, transportation and households were from 1890 onwards mainly met by mineral fuels. Commercially, natural gas developed somewhat later than coal and oil, but due to its more limited use and special transport requirements, it never came to dominate energy consumption.[1] The development of mechanized production and in the first half of the twentieth century the transition to semi-automatic production lines and mass-consumption required new energy inputs and resulted in major shifts in the composition of mineral fuel consumption. These developments in the composition of energy consumption reflect, in varying degrees of causality, the development level of an economy, industrialization, economic growth and also the expansion of domestic energy resource production.

Between 1880 and 1988, US primary energy use experienced major shifts in its composition. This change in the pattern of consumption was due to: 1) increasing per capita energy consumption; 2) expansion of the consumption of coal; 3) partial substitution of coal oil; 4) partial substitution of coal and oil by gas; 5) a long-term trend away from direct consumption of energy materials to processed and converted energy products; and 6) mechanical energy progressively derived from electricity instead of steam power. These major changes in the composition of primary energy consumption were reflected in the advancement of certain fuels, the efficiency of energy use, the development of oil consumption per unit of GNP compared to energy consumption per unit of GNP and the development of the mineral fuel producing industries.[2] However, such shifts in demand are not necessarily matched by the development of the domestic energy supply. The resource endowments and the cost of exploitation determine, among other factors, the supply conditions. The subsequent periods of expansion of coal, oil and to a lesser extent gas, were also a result of the dynamic relationship between industrial innovations and organization of production, utility of fuels and prices.

[1] Production of natural gas started around 1880 but expanded at a slower rate than oil production. Nonetheless, the share of natural gas in total energy production increased from 2.6% in 1900, to 3.2% in 1910, to 3.8% in 1920, to 9.1% in 1930, to 11.3% in 1940 and to 19.2% in 1950. The share of natural gas in total consumption increased from 2.6% in 1900, 3.2% in 1910, 3.8% in 1920, 8.1% in 1930, 10.6% in 1940 and 17% in 1950. Calculated from S.H. Schurr and B.C. Netschert, *Energy in the American economy, 1850-1975; an economic study of its history and prospects*, Greenwood Press Publishers, Westport, 1977, Table VII, Statistical Appendix to Part I, p.511-512.

[2] The analysis of price, production, consumption and economic growth over a long period of time is hampered by statistical imperfections. Some statistics are unavailable, other statistics are not comparable, and the level of aggregation of data can hide developments which are important to the analysis of the oil industry. Changes in the contributions to GNP by certain sectors influence the development of energy consumption, which cannot be derived from the energy intensity ratio. In addition to sectoral changes, the application of electricity has had a marked impact on the energy intensity. "Since GNP is measured as gross national cost, the substitution of more expensive energy increases GNP more than BTUs. For large industrial users in 1982, electric power was about eight times as expensive per BTU as was coal. The combination of the efficiency and convenience of using electricity makes it worthwhile, but the impact on the ratio of BTUs per dollar of GNP is to reduce it. In a less dramatic way, the same shift occurs for the same arithmetic reason every time anyone substitutes an expensive form of energy for a cheap one. Certainly, one of the distinctive trends in energy use during the twentieth century has been the substitution of fuels with relatively high cost per BTU for fuels with lower monetary cost. Thus wood and coal were displaced by electricity and oil. As energy prices fell relative to wage rates, consumers were willing to pay for the convenience of electricity and refined petroleum. But a shift to electric heat from coal, for example, creates a decrease in the BTU/GNP ratio, just because the electricity is so expansive per BTU." W.S. Peirce, op.cit., p.49-50.

2. Development of energy consumption

In 1880, coal contributed 41.1% to total energy consumption and oil only a mere 1.9%. Between 1880 and 1900, coal developed into the main energy input of the US economy (See Table 2.1). Oil was still mainly used as lamp oil, which limited its contribution to energy consumption. The composition of energy consumption in 1900 demonstrates the importance of coal.

Table 2.1 US total energy consumption, oil and coal consumption; 1880-1955; in trillion BTU

year	total	oil	coal
1880	5,001	96	2,054
1890	7,012	156	4,062
1900	9,587	229	6,841
1905	13,212	610	10,001
1910	16,565	1,007	12,714
1915	17,764	1,411	13,294
1920	21,378	2,634	15,504
1925	22,411	4,156	14,706
1930	23,708	5,652	13,639
1935	20,456	5,499	10,634
1940	25,235	7,487	12,535
1945	32,700	9,619	15,972
1950	35,136	12,706	12,913
1955	40,796	16,328	11,703

source: S.H. Schurr and B.C. Netschert, *Energy in the American economy, 1850-1975; an economic study of its history and prospects*, 1977, Appendix to Part I, Table VII, p.511-513.

The share of coal had rapidly increased to 71.5%, while the share of oil increased only to 2.4%. The average annual increase of coal consumption was 11.7% between 1880 and 1900, compared to an average annual increase of oil consumption of 6.9%. The expansion of coal consumption lasted until 1920, when consumption reached a level of 15,504 btu and a share in total energy consumption of 72.5%. After 1920, coal consumption declined steadily (except during World War II) and oil consumption expanded, basically at the expense of the share of coal in total energy consumption. In 1920, the share of oil was already 12.3% and increased to 23.8% in 1930 and to 29.7% in 1940. In 1950, the shares of coal and oil in total energy consumption were practically equal (36.8% and 36.2% respectively). From 1950 to 1973, the share of oil in primary energy consumption continued to increase. Coal had been substituted by oil as the major energy input in the economy. Yet, the substitution of coal was not complete, and a certain degree of diversification of energy consumption was maintained. The fact that the US was a substantive producer of all mineral fuels must be considered as the main influence on the extent of substitution.[3]

The share of oil in total energy consumption in the first half of the twentieth century did mask the significance of this particular fuel. The energy input of the newly developed and expanding manufacturing industries was principally oil and the development of road and air transport depended entirely on oil products. The expansion of oil consumption was closely linked to the expanding manufacturing industries, while coal remained the (decreasing) input of the heavy industries. Another important factor, positively influencing oil consumption, was the technical advance in techniques which improved processing efficiency and the yield of oil products. Refining techniques were until the World War I mainly based on separating the components of crude (distillation) which yielded a fairly static output of products because the product yields were entirely dependent on the quality of the available crudes. From 1900 onwards, increasing demand from the automotive and other engineering industries for liquid energy fuels for motive and heating purposes bolstered research in new techniques to meet shifting demand for lighter oil products, to reduce costs, and to add flexibility to refinery output.[4] Most innovations in refining techniques became fully operational after World War I.[5]

Growth of total energy consumption can be divided into distinct periods. Growth of energy consumption followed a pattern of expansion, stagnation, and decline in two periods of more than 50 years, with smaller rates of variation in growth in the second period than in the first. Between 1880 and 1900, the average annual increase of total energy consumption was 4.6%. Between 1900 and 1910 consumption expanded rapidly from 7,012 btu to 16,565 btu, an average annual increase of 7.8%. Growth of energy consumption was arrested when the average annual increase dropped first to 2.9% (1910-1920) and to 1.1% between 1920 and 1930. As a consequence of the recession of the 1930s, energy consumption declined between 1930 and 1935 but recovered in the second half of the decade. Average annual growth

[3] Government intervention in the mineral fuel industry did certainly influence the composition of primary energy consumption, and thus the rate of substitution.

[4] The relative shares of petroleum products as a percentage of crude input changed perceptively between 1910 and 1920. The share of illuminating oil decreased from 75.2% in 1880 to 33.0% in 1909 and decreased futher to 24.1% in 1914, 13.3% in 1918, 12.7% in 1920 and 8.1% in 1925. The share of gasoline (including naphta, benzine and aviation gasoline) developed reversely. The shares increased from 10.3% in 1880 to 10.7% in 1909. In 1914 this share had already increased to 18.2% and increased to 25.3% in 1918 and 32.4% in 1925. S.H. Schurr and B.C. Netschert, op.cit.., 1977, p.95.

[5] H.F. Williamson, R.L. Adreano, A.R. Daum, G.C. Klose, *The American Petroleum Industry, the age of energy 1899-1959*, Northwestern University Press, Evanston, Illinois, 1963, p.110-167.

Table 2.2 Indexes of energy consumption per unit of Gross National Product; 1880-1955; 1900=100

year	GNP	total energy	oil/gas	coal
1880	50.0	104.4	39.9	60.1
1885	57.3	102.7	44.2	72.4
1890	68.6	106.6	125.2	86.6
1895	81.4	98.2	80.5	88.9
1900	100	100	100	100
1905	125.3	110	162.9	116.7
1910	147.9	116.8	217.4	125.7
1915	158.2	117.1	274.8	122.8
1920	191.9	116.2	377.9	118.1
1925	237.0	98.6	479.9	90.7
1930	247.1	99.3	653.5	80.1
1935	239.4	89.1	659.5	64.9
1940	316.8	83.1	684.1	57.8
1945	437.7	72.0	613.6	49.3
1950	490.6	74.7	824.5	38.5
1955	597.6	71.2	922.9	28.6

source: S.H. Schurr and B.C. Netschert, ibid., 1977, Appendix to Part I, Table XIII, p.524-525.

between 1930 and 1940 was as a result only 0.6%.[6] Due to the American war effort, average annual growth of energy consumption rose again to 3.9%[7] between 1940 and 1950, and almost maintained this level of growth in the 1950s.[8] However, in the 1960s, growth of energy consumption accelerated to an average annual increase of 5.1%. Post-war economic growth was also buoyant, between 1950 and 1960 GNP increased at an average annual growth rate of 3.8% and 4.5% between 1960 and 1970. But after 1970, the relationship between GNP growth and growth of energy consumption transformed. In the period 1970-1988, GNP increased at an average annual rate of 3.4%, while energy consumption stagnated and grew only 13% between 1970 and 1980 and 2% between 1980 and 1988.

[6] Calculations based on Table 2.1. However, based on Table 2.3 the average annual increase of energy consumption was -0.7% between 1929 and 1940. The difference results from a deviation in the figures for 1940 between S.H. Schurr and B.C. Netschert, op.cit., 1977 and *Oil & Gas Journal Data Book*, 1989 edition, Penwell Books, Tulsa, Oklahoma, 1989.

[7] Based on *Oil & Gas Journal Data Book 1989*, the average annual increase between 1940 and 1950 was 4.4%.

[8] Between 1950 and 1960, average annual growth was 3.1%. Calculations 1950-1988 based on Table 2.3, *Oil & Gas Journal Data Book 1989*.

The relationship between energy consumption and GNP growth[9] (See Table 2.2 and 2.3) shows that, with the exception of 1895, energy consumption per unit of GNP was increasing until 1920, decreasing (except in 1933) until 1965, temporarily increasing at the end of the 1960s and beginning of the 1970s and decreasing again from 1973 to 1988.

The declining trend of the energy intensity ratio from the 1920s onwards is remarkable since it occurred in a period when real energy prices (until 1973), and at times nominal energy prices, were declining. One explanation is that the income elasticity of energy demand decreases beyond a certain level of income. At a low initial level of energy consumption, increased income will be translated into increased energy consumption, but at higher levels of income the relationship between demand and income becomes less strong.[10] The fact that energy is an intermediate good, and not a consumption good, invalidates the simple expectation that the quantity demanded increases with income. Energy is considered as an essential input into a technologically advanced economy. The substitution of labour for capital has been instrumental in the process of economic growth, and required, at first, increasing energy inputs. The technological advances that initially resulted in increasing energy inputs can, surely in part, be influenced by the availability and price of energy, and consequently decrease the input of energy per unit of output. These factors stipulate the fact that the relationship between energy consumption and GNP growth is not a simple one. In addition, there are complications in statistical measurement, especially in a period of rapid changes. Economic data are highly aggregated, usually weighted in prices, and the choice of year and the definition of how to calculate these data can result in surprisingly different outcomes.[11] In general, the findings in a variety of studies do not show any deviation from the trend. Nevertheless, some caution is required.

Energy consumption per unit of GNP increased 133% between 1880 and 1910, an average annual increase of 2.9%. The period of comparative stability between 1910 and 1920 was in fact a transition stage to a period of decline. The energy input per unit of GNP became negative in the period 1916-1920 to 1921-1925 (-2.5%).[12] Between 1920 and 1955 energy consumption per unit of GNP declined 35%, an annual average decline of 1.2% from 1920 to 1955.[13] This decline continued until the end of the 1960s (between 1950 and 1970 a percentage change of -1.4), and was followed by a short period of increase (1970-1973). After 1973, energy consumption per unit of GNP decreased again. Over the entire period

[9] See note 2.

[10] W.S. Peirce, op.cit., p.50.

[11] The energy intensity ratios of the U.S. economy presented by Peirce (p.49) are much higher than those used in this study, which are based on *The Oil & Gas Journal Data Book* 1989. The only obvious disctinction is that Peirce's data are calculated in $1972 GNP and *The Oil & Gas Data Book*'s in $1982 GNP. The observed trend is similar.

[12] S.H. Schurr and B.C. Netschert, op.cit., p.163.

[13] S.H. Schurr and B.C. Netschert, op.cit, p.163.

Table 2.3 Energy and oil consumption and efficiency 1929-1988; GNP in 1982 $

year	GNP billion 1982 $	energy cons, trillion BTU	energy cons per GNP$, 1,000 BTU	oil cons, trillion BTU	oil cons per GNP$ 1,000 BTU
1929	711.8	23,756	33.4	5,521	7.8
1933	501.6	16,900	33.7	4,833	9.6
1940	775.0	23,589	30.4	7,407	9.6
1946	1,077.8	30,494	28.3	10,002	9.3
1950	1,206.6	33,992	28.2	13,489	11.2
1955	1,494.9	39,703	26.6	17,524	11.7
1960	1,665.3	44,569	26.8	20,067	12.1
1965	2,087.6	53,343	25.6	23,242	11.1
1970	2,416.2	67,143	27.8	29,537	12.2
1971	2,484.8	68,348	27.5	30,570	12.3
1972	2,608.5	71,643	27.5	32,966	12.6
1973	2,744.1	74,282	27.1	34,840	12.7
1974	2,729.3	72,543	26.6	33,455	12.3
1975	2,695.0	70,546	26.2	32,731	12.1
1976	2,826.7	74,362	26.3	35,175	12.4
1977	2,958.6	76,289	25.8	37,122	12.5
1978	3,115.2	78,088	25.1	37,965	12.2
1979	3,192.4	78,898	24.7	37,123	11.6
1980	3,187.1	75,952	23.8	34,202	10.7
1981	3,248.8	73,989	22.8	31,931	9.8
1982	3,166.0	70,840	22.4	30,232	9.5
1983	3,279.1	70,497	21.5	30,054	9.2
1984	3,501.4	74,060	21.2	31,051	8.9
1985	3,607.5	73,944	20.5	30,922	8.6
1986	3,713.3	74,260	20.0	32,196	8.7
1987	3,821.1	76,042	19.9	32,627	8.5
1988*	3,932.3	77,500	19.7	33,290	8.5

source: *Oil & Gas Journal Data Book*, 1989 edition, p.112; *1988: estimate supplied by *Oil & Gas Data Book*.

1950-1988, energy consumption per unit of GNP declined[14], except for the short rise at the end of the 1960s and beginning of the 1970s. In this period, the American economy was experiencing a fundamental disequilibrium, but the necessary economic measures were delayed because of domestic political problems, the American involvement in the VietNam war and the crisis in international monetary relations.[15] American GNP increased 15.5% between 1965 and 1970, while energy consumption increased 25.9% and oil consumption increased 27.1%. Declining real oil prices (See Figure 2.1) in the 1960s combined with an over-valued dollar were another incentive for the expansion of US oil consumption.

The development of coal consumption per unit of GNP was quite similar to the development of energy consumption per unit of GNP in the beginning of the twentieth century. In fact, the periods of growth, stagnation and decline of consumption per unit GNP corresponded closely. From 1880 to 1905 coal consumption per unit of GNP was increasing, remaining more or less stable between 1905 and 1920, and then declining after 1920.

Oil consumption per unit of GNP, however, developed in a distinctive way from both coal consumption and total energy consumption per unit of GNP and increased, at a varying pace, from 1900 until 1973. Thus, the period of rising energy consumption per unit of GNP corresponds basically with the period of increasing coal consumption, while the period of decreasing energy consumption per unit of GNP corresponds, until 1973, with the increased consumption of oil and to a lesser extent gas.

The percentage change in oil consumption per unit of GNP (See Table 2.4) was 23.1 from 1929 to 1940. Between 1940 and 1950, the percentage change was 16.7 and 8.9 between 1950 and 1970. In the period 1970-1987, the percentage change in oil consumption per unit GNP was -30.3. This clearly shows a diminishing relationship between GNP growth and oil consumption, and indicates the maturation of the American oil market.

Oil consumption per unit of GNP increased annually from 1880 (except in 1895, 1946, 1965) until 1973. In the 1970s, oil consumption per unit GNP fluctuated. It dropped from 12.7 in 1973 to 12.3 and 12.1 in 1974 and 1975 respectively, increased again in 1976 (12.4) and in 1977 (12.5), then decreased from 1978 (12.2) to 1985 (8.6), increased in 1986 (8.7), decreased again in 1987 (8.5) and remained stagnant in 1988.

In the period 1970-1980, oil consumption per unit of GNP declined at a faster rate than energy consumption per unit of GNP. Oil consumption per unit of GNP peaked in 1973, declined slightly between 1973 and 1978 and at a faster rate after 1978. The relationship between the oil price increases of 1973/1974 and 1979/1980 and the diminishing oil/GNP

[14] Energy consumption per unit of GNP was: 1940-1950: -7.2; 1950-1970: -1.4; 1970-1987: -28.4. *Oil & Gas Journal Data Book 1989*, p.112.

[15] "By 1968, the inflation generated by the war had become counterproductive to the overall American economy. The most powerful U.S. industrialists saw the writing on the wall in the form of monetary crisis, payments deficits, blocks on capital flow, a growing hostility to their investments in Europe, competitive loss of markets, and especially the rapid loss of value of the dollar." J. Kolko, *America and the crisis of world capitalism*, Beacon Press, Boston, 1974, p.6.

ratio is obvious. Real oil prices declined between 1974 and 1978, and after the initial effect of the price increases, the oil/GNP ratio increased again in 1976 and 1977. The oil price increase of 1979/1980 had more lasting effects on the oil/GNP ratio. Government diversification and energy efficiency policies were more actively pursued, and the private sector expedited to diminish energy costs per unit of output. The oil/GNP ratio declined to 8.6 in 1985, recovered to 8.7 in 1986 when oil prices dropped below $10 a barrel and decreased to 8.5 in 1987 and 1988.

Table 2.4 Percentage change of energy consumption and efficiency

years	GNP billion 1982$	energy cons trillion BTU	energy cons per GNP$, 1,000 BTU	oil cons trillion BTU	oil cons per GNP$, 1,000 BTU
1929-1940	8.9%	-0.7%	-9.0%	34.2%	23.1%
1940-1950	55.7%	44.1%	-7.2%	82.1%	16.7%
1950-1970	100.2%	97.5%	-1.4%	119.0%	8.9%
1970-1987	58.1%	13.3%	-28.4%	10.5%	-30.3%

source: *Oil & Gas Journal Data Book*, 1989 edition, p.112.

The pattern of oil consumption growth can be divided into distinct periods of expansion and stagnation. Oil consumption (See Table 2.6) experienced between 1860 and 1880 high average annual growth rates (1860-1870: 30.2% and 1870-1880:75.5%), moderate growth rates between 1880 and 1900 (1880-1890: 6.1% and 1890-1900: 4.3%) and went through a period of major expansion between 1900 and 1930 (1900-1910: 33.7%; 1910-1920: 16.2% and 1920-1930: 11.4%). The expansion of oil consumption in this period was also displayed in the increase of per capita oil consumption. Per capita petroleum consumption increased from 0.02 in 1860 to 0.14 in 1870, 0.52 in 1880, 0.73 in 1890, 0.85 in 1900, 2.28 in 1910, 4.31 in 1920, 7.53 in 1930 and 8.51 in 1936.[16] Between 1930 and 1940, the average annual growth rate of oil consumption was a moderate 3.2%. Oil consumption between 1940 and 1950 increased to an average annual growth rate of 7.0% and 4.9% in the two following decades. Between 1970 and 1980, the average annual increase in consumption was only 1.5% and declined -0.02% between 1980 and 1988.

3. Development of oil production

The dominance of coal consumption between 1880 and 1940 was also expressed in the share of coal production in total energy production (See Table 2.5). Coal's share in total US energy

[16] In 1936 per capita petroleum consumption in other countries was still lagging far behind the United States. Canada ranked 2nd with 3.68, the United Kingdom 8th with 1.71 and the Netherlands 15th with 1.07. *Petroleum Facts and Figures*, 1937, Department of Public Relations, American Petroleum Institute, New York, p.7.

production increased from 40.7% in 1880 to 76.4% in 1900. Until 1930, it had remained rather stable at around 75%, but from 1930 onwards it declined to 40.1% in 1950. Meanwhile, the share of oil production in total energy production increased from 3.7% in 1900 to 22.% in 1930 and to 32.1% in 1950. Unlike the massive shift in the composition of consumption, coal maintained a relatively large share in energy prodution. In 1988, 465.8 million tonnes o.e. of coal, 462.5 million tons of oil, and 425.7 million tonnes o.e of gas were produced.[17]

Table 2.5 US total energy production, oil and coal production 1880-1955; in trillion BTU

year	total	oil	coal
1880	5,055	146	2,058
1890	7,148	258	4,096
1900	9,908	369	7,020
1905	13,386	781	10,228
1910	17,140	1,215	13,074
1915	18,510	1,630	13,857
1920	22,925	2,569	17,175
1925	23,140	4,430	15,195
1930	23,119	5,208	14,011
1935	21,224	5,780	11,081
1940	26,446	7,849	13,380
1945	33,594	9,939	16,529
1950	35,672	11,447	14,647
1955	39,967	14,410	12,839

source: S.H. Schurr and B.C. Netschert, op.cit., 1977, Appendix to part I, Table II, p.495-499.

The oil industry experienced a period of rapid expansion after the discovery of the Drake well in 1859 (See Table 2.6). Oil production increased at an average annual growth rate of 95.2% between 1860 and 1870 and oil consumption at a rate of 30.2%. From 1870 to 1880, production grew at a slower average annual rate than consumption, 39.9% and 75.5% respectively. Oil prices decreased in this era from $9.59 a barrel in 1860 to $3.86 in 1870 and $0.95 in 1880 (See Figure 2.1). The two decades of major expansion and competition, were followed by two decades of moderate average annual growth of consumption and production, and concentration. From 1880 to 1900, consumption grew at a slower rate than production, roughly at a ratio of 1.0 to 0.6.[18] In this period of moderate growth of

[17] *BPSR 1989*, p.4, 22, and 26.

[18] S.H. Schurr and B.C. Netschert, op.cit., p.88.

production and consumption, the oil industry was dominated by the Standard Oil Trust.

The market for lamp oil had given the oil industry its impetus, and initiated the first crude oil growth cycle. At its inception, this market was competitive in both the United States and in the rest of the world. Companies supplied the Eastern markets with oil from Russia and later Indonesia. The American industry supplied the domestic market with lamp oil and exported up to a third of production to Europe for the remainder of the century.[19] Oil prices declined from over $9 a barrel nominally to less than a $1 between 1860 and 1880 (See Figure 2.1). Nevertheless, the margins on refining were strained, because refining capacity had been expanding faster than the supply of crude.[20] Also the transportation sector was suffering from overcapacity relative to available traffic. The oil industry was as a result highly unstable because its small specialized firms had high unit costs and low profit margins, which offered little flexibility in difficult times.[21] This instability was aggravated by the uncertainty of finding oil and the exploitation of oil wells under the ownership rules in the US. In order to secure better market conditions in refining, Rockefeller of Standard Oil, started to strengthen his market position at the end of the 1870s through the acquisition of existing refineries, the conclusion of transport contracts, and the expansion of his distribution network in the US and the international market.[22] To illustrate the strength of the Standard Oil Trust, a Report of the Commissioner of Corporations on the Petroleum Industry of 1907 states: "In the year 1904 the Standard Oil Company and affiliated concerns refined over 84 per cent of the crude oil run through refineries; produced more than 86 per cent of the country's total output of illuminating oil; maintained a similar proportion of the export trade in illuminating oil; transported through pipe lines nearly nine-tenths of the crude oil of the older fields and 98 per cent of the crude of the Mid-Continent, or Kansas-Territory field; secured over 88 per cent of the sales of illuminating oil to retail dealers through the country, and obtained in certain large sections as high as 99 per cent of such sales. It also controlled practically similar proportions of the production and marketing of gasoline and lubricating oils."[23]

[19] C. Tugendhat and A. Hamilton, *Oil, the biggest business*, Eyre Methuen, London, 1975, p.11.

[20] H.M. Larson, The rise of Big Business in the Oil Industry, in: *Oil's First Century*, Harvard Business School, Papers Given at the Centennial Seminar on the History of the Petroleum Industry, November 13-14, 1959, p.31.

[21] Ibid., p.31.

[22] "While the industry was made up of hundreds of small units there was no possibility of earning a consistent profit. On the one hand, the price of crude oil fluctuated wildly, depending on the level of production in the regions and the number of new wells that were being drilled, while, on the other hand, refiners had to carry on cut-throat competition with each other to win contracts from the retailers. Rockefeller estimated that in 1871 'more than three-fouths of the oil refiners of the country did a losing business' and he saw no hope of the situation improving until they came together." C. Tugendhat and A. Hamilton, op.cit., p.16-17.

[23] I. Skeet (ed.), op.cit., p.35.

Figure 2.1

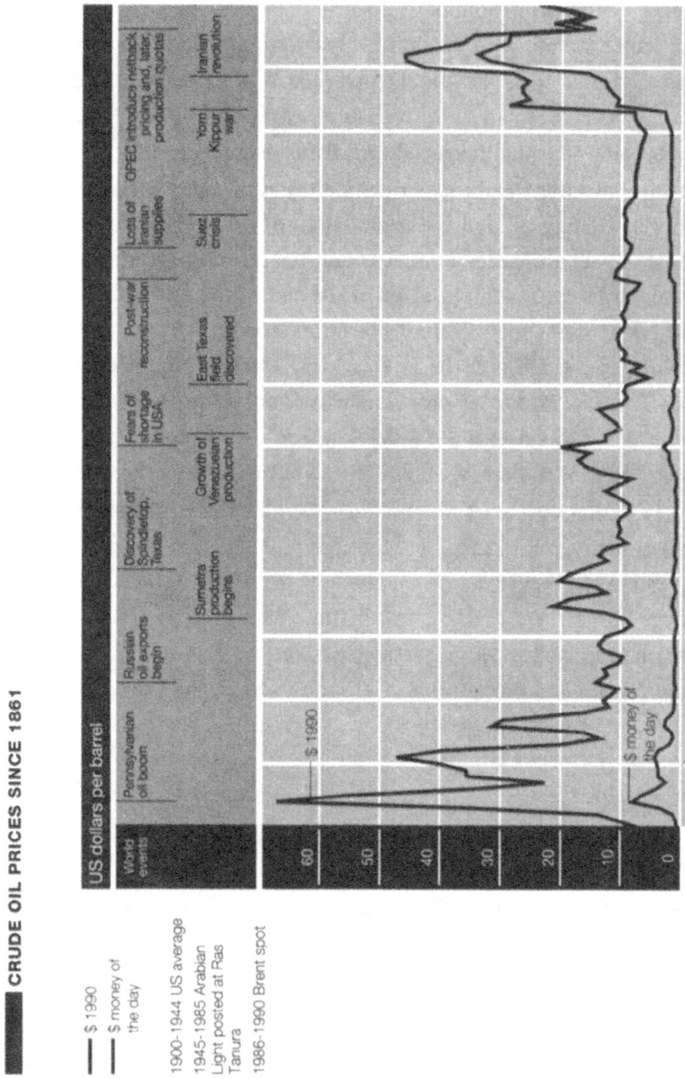

Source: BP Statistical Review of World Energy.

Table 2.6 Average annual growth rates of US consumption and production 1860-1990

years	consumption	production	c>p
1860-1870	30.2	95.2	-
1870-1880	75.5	39.9	+
1880-1890	6.1	7.4	-
1890-1900	4.3	3.9	+
1900-1910	33.7	22.9	+
1910-1920	16.2	11.1	+
1920-1930	11.4	10.3	+
1930-1940	3.2	5.1	-
1940-1950	7.0	4.6	+
1950-1960	4.9	3.0	+
1960-1970	4.9	3.7	+
1970-1980	1.5	0.6	+
1980-1990	-0.2	-1.3	+

source:calculated from: 1860-1950: S.H. Schurr and B.C. Netschert, op.cit., 1977, Table 22, p.86; 1950-1990: Table 2.7.

During the period from approximately 1880 untill 1900, when Rockefeller's Standard Oil Trust monopolized the US market, the average annual increase in oil consumption was less impressive than in the previous two decades, 6.1% between 1880 and 1890 and 4.3% between 1890 and 1900 (See Table 2.6), indicating the maturity of the market for illuminating oil. Production also experienced lower annual increases than between 1860 and 1880, 7.4% between 1880 and 1890 and 3.9% between 1890 and 1900. Between 1880 and 1890, production still increased at a slightly higher rate than consumption, but between 1890 and 1900, the average annual growth rate of production had become smaller than this rate of consumption. Consequently, the oil price was, except for 1883, lower than $1 a barrel throughout the 1880s and first half of 1890s, fluctuated between $0.79 and $1.36 in the second half of the 1890s, and was $1.19 in 1900.[24]

At the turn of the century, the strong market position of the Standard Oil Trust was contested from various sides. The market for lamp oil had stagnated and then faltered when electric light became available as a substitute. The second attack on Standard's position came from the Texas- and California-based oil companies. In 1901, the Spindletop was discovered and drew many new entrants into the industry. These new companies also managed to capture

[24] A. Hamilton, *Oil, the price of power*, Michael Joseph/Rainbird, London, 1986, p.9.

market shares in the newly developing lubricant market.[25] Standard Oil was handicapped in the manufacture of new oil products because of its heavy commitment to kerosene and the fact that it did not have sufficient light crudes to increase the output in gasoline. Larson observed that: "Standard was suffering from the fate that has normally overtaken pioneers in industrial development, that of being tied by earlier investments and operations."[26] The competitors of Standard Oil were thus able to build up strong positions in the rapidly expanding gasoline market. Furthermore, the geographic expansion of production within the US made the Trust's control of the market much more complicated. Especially, the Texas-based companies were able to to encroach on the Trust's traditional markets. In 1909, Standard Oil's share in US refining capacity had been reduced to 66%. The third blow to the Trust was the start in 1904 of legal actions by various governments of states demanding dissolvement of the Trust. In May 1911 the Supreme Court ruled that the Trust should be dissolved within six months.

The industrial development in the US from 1900 onwards had brought about a change in the consumption pattern of primary energy[27], which also affected the dominating market structure. Successful exploration by new entrants in the market led to an increase of oil production and subsequently competition. The official dissolution of the Standard Oil Trust in 1911 merely emphasized the drive towards more competition already initiated by market forces.

From 1900 until 1930, the average annual growth rate of oil consumption exceeded the rate of production. The increase in consumption growth was a result of a change in the use of oil. Until 1900, lamp oil was the most important product, and had driven the course of the crude oil growth cycle, while after 1900, oil was increasingly used as a fuel for internal combustion engines and heating, and this initiated a new crude oil growth cycle.

Between 1930 and 1940, growth in production exceeded growth of consumption, although production was from 1928 onwards prorationed per deposit.[28] The economic recession had affected growth of energy consumption severely. Between 1930 and 1935, oil consumption declined, while oil production still grew moderately. Growth of oil consumption, at an average annual increase of 7%, was restored in the period 1940-1950 (See Table 2.1), while the average annual increase of oil production started to diminish. The

[25] "The growth of these new markets gave the new companies an opportunity to outflank Standard's entrenched positions, and to expand far more quickly than if kerosene had remained the principal product." C. Tugendhat and A. Hamilton, op.cit., p.39.

[26] H. Larson, op.cit., in: *Oil's First Century*, p.40-41.

[27] The demand pattern in Western Europe changed less rapidly than in the United States because the innovation in industrial production took place at a later date. Coal remained the main energy input in industry until after the second world war. West Germany, France, Belgium, Holland and the United Kingdom were also major coal producing countries and, more important, had no economically viable oil producing resources. The share of coal in primary energy use in Western Europe was 83% in 1929 and still 77% in 1938. *Kolen en Energiepolitiek*, studie-comite van de Westeuropese steenkolenproducenten, Brussel, juli 1955, p.13.

[28] See section 6.

expansion of American oil production subsided, and the industry reached maturity in the decades after World War II. Average annual growth of production ultimately became negative in the period 1980-1988. From 1947 onwards, domestic production could not satisfy consumption of oil, and the US increasingly became a net importer of oil (See Table 2.7).

From 1947 to 1959, the year in which the US government introduced an import quota programme[29], self-suffiency in oil declined rapidly. Net imports as a percentage of consumption increased from 5.5% in 1949 to 16.7% in 1959. The import control programme could not, however, arrest the declining self-sufficiency and the share of net imports in consumption increased throughout the 1960s. The development of the Alaska oil fields (discovered in 1968) and other off-shore developments was stalled, when an off-shore oil field caused substantial ecological damage in 1969. President Nixon abolished the ineffective import programme in April 1973. The share of net imports in consumption had in the mean-time increased to 35.7%. The share of net imports in consumption continued to grow until 1977 (47.2%), when the Alaska oil production finally came on stream. A combination of decreased oil consumption and increased domestic oil production diminshed the share of net imports in consumption to 28.6% in 1982. From 1984, oil consumption increased again while domestic oil production was stagnating and decreasing from 1987 onwards. As a result, the share of net imports in consumption started to increase again and reached 36.9% in 1988.

4. Prices; 1880-1940

The development of nominal oil prices from 1860 to 1973 (See Figure 2.1) displays a relative stability. From 1872 onwards, nominal oil prices exceeded $1.50 a barrel only in the periods 1917-1926 and 1947-1961, both after-war periods.[30] Real oil prices display a long-term decline from 1859 to 1973. Once more, the real oil price is in the current period approaching its pre-1973 level. This would suggest that the developments between 1973-1985 were an aberration in the long-term development of real oil prices. Obviously, both the nominal and real oil price paths do not comply with Hotelling's price path, which predicted oil prices to increase at the rate of interest. Griffin and Teece[31] have argued that the producer's expectations about present and future demand and supply conditions- which are influenced by the expectation with regard to the size of the reserve base, the availability of a backstop fuel, the rate of discount, price elasticities of demand, and world economic growth- would suggest that "the price of oil is likely to be highly unstable, with the effect of switching to

[29] See Chapter 4.

[30] Oil prices in Figure 2.1 are: from 1859-1899 Pennsylvania Crude prices; from 1900-1940 US Average Crude prices; 1945-1973 Arabian Light posted prices; and 1973-1985 Arabian Light Official Goverment price. From 1957 to 1972 market prices of crude oils were at a discount on posted prices and ranged between $1.3-$1.5 a barrel in the period 1961-1970.

[31] J.M. Griffin and D.J. Teece, op.cit., in: *OPEC Behavior and World Oil Prices*.

Table 2.7 US oil production, consumption, trade balance, and net imports as percentage of consumption; 1938-1990; in thousand barrels per day

year	consumption	production	imports	exports	net-imp % of consumption
1938	3,120	3,330	130	500	-11.9
1949	5,800	5,050	650	330	5.5
1950	6,500	5,410	850	310	8.3
1951	7,040	6,160	840	420	6.0
1952	7,280	6,260	960	430	7.3
1953	7,600	6,460	1,030	400	8.3
1954	7,760	6,340	1,050	360	8.9
1955	8,460	6,810	1,250	370	10.4
1956	8,780	7,170	1,450	430	11.6
1957	8,820	7,170	1,570	570	11.3
1958	9,080	6,710	1,700	280	15.6
1959	9,420	7,055	1,780	210	16.7
1960	9,660	7,035	1,820	210	16.7
1961	9,810	7,185	1,890	180	17.4
1962	10,230	7,330	2,120	170	19.1
1963	10,550	7,540	2,130	210	18.2
1964	10,820	7,615	2,260	200	19.0
1965	11,300	7,805	2,470	190	20.2
1966	11,850	8,295	2,540	200	19.7
1967	12,275	8,810	2,540	310	18.2
1968	13,085	9,095	2,810	200	19.9
1969	13,815	9,240	3,170	230	21.3
1970	14,350	9,635	3,420	260	22.0
1971	14,845	9,465	3,930	220	25.0
1972	15,990	9,440	4,740	225	28.2
1973	16,870	9,210	6,255	230	35.7
1974	16,150	8,795	6,125	220	36.6
1975	15,875	8,375	6,025	210	36.6
1976	16,980	9,735	7,290	220	41.6
1977	17,925	9,865	8,710	250	47.2
1978	18,255	10,275	8,225	365	43.1
1979	17,910	10,135	8,410	485	44.2
1980	16,460	10,170	6,735	555	37.5
1981	15,550	10,180	5,950	605	34.4
1982	14,765	10,200	5,040	815	28.6
1983	14,745	10,245	4,990	740	28.8
1984	15,175	10,505	5,380	720	30.7
1985	15,170	10,545	5,065	780	28.2
1986	15,560	10,235	6,045	765	33.9
1987	15,955	9,945	6,245	745	34.5
1988	16,420	9,750	6,750	694	36.9
1989	16,595	9,160	8,019	817	43.4
1990	16,200	8,850	8,026	889	44.1

Source: BPSR 1959, 1967, 1977, 1987, 1989, 1991.

alternative price paths possibly dominating the more systematic movements along a given price path."[32] The short-term oil price movements are indeed changeable, but the long-term nominal oil price developments show a larger degree of stability. The downward trend in real oil prices from 1860 to 1970, however, and the impressive increase of real oil prices in 1973 and 1980, followed by a steep decrease in 1985, casts doubt upon the effectiveness of market regulation by the market participants over longer time periods.

From 1859 to 1880, nominal oil prices were erratic but were, in the longer-term, declining (See Figure 2.1). The declining oil prices in the 1870s occurred in a period of rapid expansion of the industry. Declining prices, deceleration of growth of demand towards the end of the decade, and over-capacity in refining made price competition less atractive. The producers exhibited typical market behaviour of producers of a product in a maturing stage. The decade was characterized by cartels and takeovers, and the Standard Oil Trust was formed in this period. The increase in concentration of the industry could not arrest the decline in prices in the 1880s, but the price variations were already less pronounced than in the decade before.[33] In the 1890s, oil prices increased again and were $1.19 in 1900. New companies entered the industry in the 1890s and the beginning of the twentieth century, both as a result of higher prices and changing demand conditions. The discovery of the Spindletop in 1901 resulted in a drop of the oil price below $1 a barrel[34] and only in 1916 did the price of oil increase above this level ($1.10 a barrel). During this period (1900-1915), the lamp oil market went into decline, but the fuel market experienced rapid expansion. The market structure in the US and abroad was competitive.

As a result of war-time inflation[35] and a tightening of supply, oil prices increased after 1915 and reached $3.07 in 1920. The oil price increases in the 1920s exceeded the rise in wholesale prices, and this continued until 1929. Comparison of the development of the index crude oil/wholesale prices and the index coal/wholesale prices (See Table 2.8) reveals that both coal and oil price indexes increased less than the wholesale price index from 1900 to 1920-1925. The coal/wholesale price index continues to increase more than the wholesale price index from 1925 to 1950, while the oil price/wholesale price index indicates that between 1935 and 1945, the crude oil price index increased less than the wholesale price index. A comparison of US retail price indexes of gasoline with the indexes of foods,

[32] Ibid, p.24.

[33] A. Clo, Storia e cronaca del prezzo del petrolio, in: *Energia*, 2-1986.

[34] In 1901, oil prices per barrel at the well were $0.18 for Texas Gulf Coast, $1.07 for Rocky Mountain, $0.85 for Lima-Indiana, $5.00 for Indiana Illinois, $0.57 in California, $0.79 in the Mid Continent and an average of $1.21 for Apalachian crudes. In 1899, Gulf Coast crudes averaged $5.00 a barrel. The discovery of the Spindletop and the subsequent production in 1901 depressed the average US oil price. *Petroleum Facts and Figures*, 1937, fifth edition, American Petroleum Institute, New York, p.92 and 96.

[35] Between 1915 and 1920 oil prices increased yearly as a result of World War I. The crude oil prices per barrel were in 1916 $1.10, in 1917 $1.56, in 1918 $1.98, in 1919 $2.01. The index of oil prices relative to the wholesale price index was 98.8 in 1916, 101.9 in 1917, 115.9 in 1918 and 111.3 in 1919. S.H. Schurr and B.C. Netschert, op.cit., Appendix to part I, Tables XXIII and XXIV, p.545-548.

housing, clothing, fuel and light and sundries, shows that gasoline prices (exclusive of tax) also declined more between 1926 and 1936 than prices of the other commodities.[36]

Table 2.8 Comparison of coal, oil and gas prices and price indexes; 1880-1955; 1947=100

year	index oil /whlsale	oil price $	index	gas price in $cents	index	coal price $	index	index coal /whlsale
1880	-	0.94	48.7	-	-	1.47	20.4	-
1885	-	0.88	45.6	7.0	116.7	2.00	27.7	-
1890	-	0.77	39.9	7.0	116.7	1.43	19.8	-
1895	-	1.09	56.5	7.0	116.7	1.41	19.5	-
1900	162.8	1.19	61.7	7.1	118.3	1.49	20.6	54.4
1905	79.1	0.62	32.1	5.0	83.3	1.83	25.3	62.3
1910	66.5	0.61	31.5	5.3	88.3	1.90	26.3	55.4
1915	70.8	0.64	33.2	6.1	101.7	2.07	28.7	61.2
1920	152.6	3.07	159.1	9.4	156.7	4.85	67.6	64.6
1925	124.6	1.68	87.0	9.4	156.7	5.30	73.4	105.2
1930	106.0	1.19	61.7	7.6	126.7	5.11	70.8	121.6
1935	93.3	0.97	50.3	5.8	96.7	4.03	55.8	103.5
1940	99.6	1.02	52.8	4.5	75.0	3.99	55.3	104.3
1945	88.5	1.22	88.5	4.9	81.7	5.90	81.7	114.4
1947	100	1.93	100	6.0	100	7.22	100	100
1950	121.6	2.51	130.1	6.5	108.3	8.90	123.3	115.2
1955	125	2.77	143.5	10.4	173.3	4.50	108.2	94.9

crude oil: $ per barrel; natural gas dollar cents per cubic feet; and coal $ per net ton. source: S.H. Schurr and B.C. Netschert, op.cit., Appendix to part I, Tables XXIII and XXIV, p.545-548.

The real price of oil also exhibits a substantial increase (See Figure 2.1) between 1915 and 1920, but it declined sharply in the period 1925-1930. From 1935 to 1940, real oil prices rose once again. As a result of the oil price controls, imposed during the Second World War, the price of oil declined substantially between 1940 and 1945.[37]

By the 1920s, expansion of supply and demand was already diminishing as compared to the two earlier decades, and oil prices began to decline from 1927 onwards. In the 1920s, both domestic and international competition was strong. The governments of Great Britain

[36] Prices of other oil products like kerosene, fuel oil and lubrcating oil were all lower in 1936 than in 1926. *Petroleum Facts and Figures*, 1937, p.158.

[37] During World War II prices did not change very much, in five years 8 dollar cents and the index relative to wholesale price index, in fact, dropped from 110.3 in 1941 to 92.6 in 1942, to 89.5 in 1943 and 89.4 in 1944. S.H. Schurr and B.C. Netschert, op.cit., Appendix to part I, Tables XXIII and XXIV, p.545-548.

and France secured their interests in oil at the San Remo Conference, much to the dislike of the American government. In the early 1920s, a general concern developed in the US about the self-sufficiency in oil. As a result, the US government interfered in the negotiations between the European governments and companies to secure part of the lucrative Middle East oil concessions for American oil companies.[38] But towards the end of the 1920s, when these interests in the Middle East had been secured, the oil market was flushed instead of short of supply. In 1928, the American government implemented a demand prorationing system and three large oil companies (Exxon, BP and Shell), operating on the international oil market, entered into a cartel agreement to regulate demand and supply.

The economic recession aggravated the industry's declining market cycle and the industry went through a difficult period. Between 1930 and 1940, the average growth of consumption was exceeded by the average growth of production and oil prices declined to $0.67 in 1933. In the second half of the 1930s, oil prices restored slightly and surpassed the $1 a barrel level again.

5. The dominance of the US oil industry in the world; 1880-1940

The expansion of world oil production between 1860 and 1940 (See Figure 2.2 and 2.3) was predominantly driven by the expansion of US oil production. In 1880, there were 7 oil producing countries in the world: Romania (0.3%), the US (87.6%), Italy (0.007%), Canada (1.2%), the Soviet Union (10.0%) and Japan and Taiwan (0.03%).[39] In 1890, oil was also produced in India (0.15%). In 1890, the share of the US in world oil production had decreased to 57.8% and the Soviet Union's had increased to 37.4%. By 1900, the Soviet Union (50.8%) had surpassed the US (42.7%) as the biggest producer. Shortly after the Spindletop came on stream, the US surpassed the Soviet Union again as the world's biggest producer.

The pattern of oil production in the world changed further in favour of the US when the Soviet Union began to suffer production problems. In 1910, production in the Soviet Union was lower than in 1900 (70,337 thousand barrels compared to 75,780 thousand barrels). Consequently, the share of the US increased to 63.9% in that period. In 1920, production in the Soviet Union had fallen to 25,430 thousand barrels, and decreasing its share to only 3.7% of world production. The share of the US continued to be high. Mexico was then the second largest producer with a share of 22.8%. In 1930, the US produced 63.6% of world production, the Soviet Union (recovered from the revolution and subsequent civil war) 8.9%, Mexico only a mere 2.8%, after water had seeped into the most important oil wells, and Venezuela 9.8%. By 1936, 27 countries were producing oil, with the US still ranking first (60.7%), the Soviet Union ranking second (11.9%) and Venezuela ranking third (9.1%).

[38] See section 6.

[39] The shares in world production are between parentheses. *Petroleum Facts and Figures*, 1937, p.56-57.

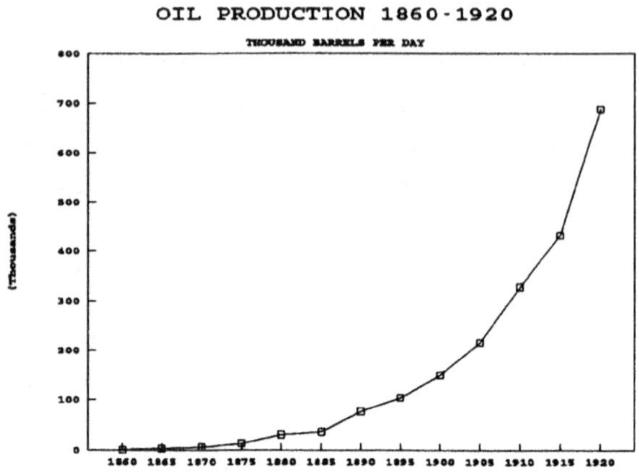

Figure 2.2

In terms of consumption, in 1936 the share of the US was almost equal to its share in production (62.0%). Europe ranked second as consumer (23.2%), while the rest of the world consumed the remaining 14.8%. The Unites States was by far the largest consumer of motor fuel (71.4%), while Europe consumed 18.6% of world motor fuel consumption. In 1936, Europe was the largest consumer of kerosene (41.2%). The US represented 38.5% of world kerosene consumption but 57.2% of world fuel oil consumption as against a share of 23.7% for Europe.

The US market was very attractive for the companies because production and consumption were physically close. Due to the expanding market, refining and transport of oil through pipelines, and/or by rail could enjoy economies of scale. Compared to other manufacturing industries, the index of production of refineries grew at a faster rate than the index of all other manufacturing industries in the period 1925-1936.

In fact, in 1937 the share of the US in world refinery capacity was 60.2%[40] and the US was also the world's most important exporter of processed oil. Compared to the shares of the other areas in the world, the US refining industry was vast. The shares of the other areas in world refining capacity in 1937 were: Western Europe 4.7%, the USSR and Eastern Europe 12.6%, Latin America and the Caribbean 12.5%, Asia 7.6% and Africa 0.1%.[41]

In the first two decades of the twentieth century, the international oil industry was, compared to the American industry, at a disadvantage. The main consuming centres were

[40] J. Darmstadter with P.D. Teitelbaum and J.G. Polach, *Energy in the World Economy, a statistical review of trends in output, trade and consumption since 1925*, Johns Hopkins Press, Baltimore and London, 1971, p.155.

[41] Ibid., p.155.

distant from the main producing centres and the markets were still fairly small, both in terms of volume and the range of oil products consumed. Non-US oil production mainly served the regional eastern hemisphere markets, and the refining industry operated at the producing centres in order to capture some economies of scale. The oil products were shipped from the source refineries to the consumer markets. Although the extraction costs were lower outside the US, the costs of transport were much higher.[42]

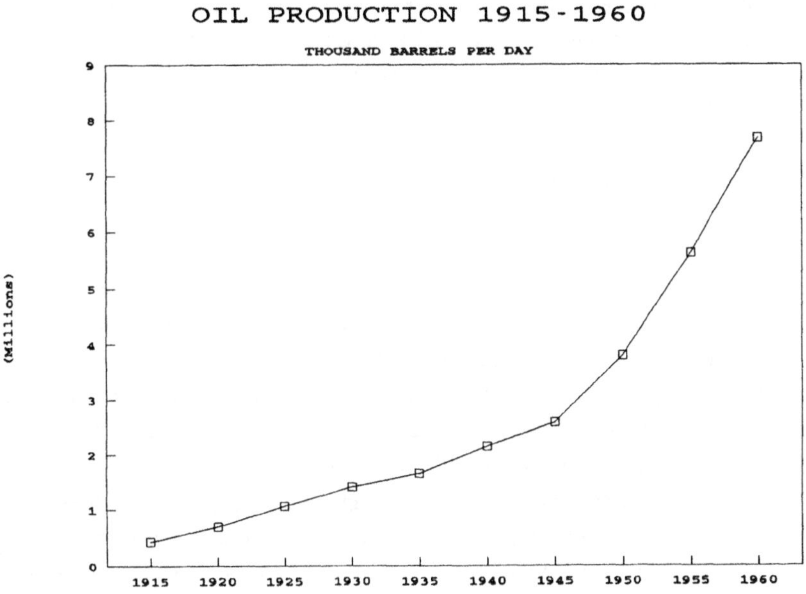

Figure 2.3

The international oil industry did, however, expand rapidly after the World War I, when new investment opportunities occurred in the Middle East. The Soviet oil production had collapsed after the revolution in 1917 and the following civil war. The Soviet Union had expropriated the oil industry, and due to the disruptions in the economy, stopped being a major exporter for some decades.[43] Shell, for example, suffered a severe loss of security

[42] The refining industry at the producing centres could not capture the full economies of scale at their inception because of limited demand for a wider spectrum of oil products. Only when demand for a variety of oil products expanded in the international market, did this industry become more competitive. However, by that time the refining industry was transferred to the main consuming centres to economize on transport costs.

[43] "With the collapse of the Russian industry in the 1917 Revolution 65 per cent of the world's production lay within the borders of the United States." C. Tugendhat and A. Hamilton, op.cit., p.74.

of supply, and the company has been a crude-short company ever since.[44] The (mainly European) companies involved in the Soviet oil industry were responding to the loss of their oil production by rerouting their investments to the Middle East. The interest of the oil companies in the Middle East was accompanied by the interest of governments of Great Britain and France which, in the absence of domestic resources, wanted to secure the oil producing potency of the Middle East. The use by the US of cars, trucks, diesel vessels and airplanes had proved their value as strategic warfare assets. The prospects in the Arabian/Persian Gulf region at that time, especially in Iraq, were very promising.[45] The outcome of the San Remo Conference in 1922 gave Great Britain and France the legitimacy to develop the region's oil potency. Negotiations to establish the Iraq Petroleum Company began in the 1920s.

6. Government intervention in the 1920s

In terms of coal consumption (See Table 2.1), total energy consumption per unit of GNP, coal consumption per unit of GNP (See Table 2.2), coal production (See Table 2.5) and oil (and gas) prices (See Table 2.8), 1920 stands out as an extraordinary year. The average annual growth rate of consumption between 1910 and 1920 was exceeding the rate of production substantially. Domestic demand for oil had almost doubled between 1915 and 1920. Between 1915 and 1920, net imports of crude oil increased from 14,371 thousand barrels a day to 96,880 thousand barrels a day, while net exports of oil products were 67,634 thousand barrels a day in 1920, creating a first time crude and oil product trade deficit in the US.[46] The recoverable crude oil reserves by year of discovery dropped sharply after 1920 and the yearly discoveries neared production. The ratios of annual new discoveries to annual production (See Table 2.9) were diminishing in the 1920s, and were substantially lower than in the previous decade[47].

Oil exploration and production in the US caused, however, serious market misallocations because of the fugitive nature of oil, the uncertain tenure of the right to receive rents and

[44] "At the time of the Revolution in 1917 Russia was producing about 15 percent of the world's oil; a third of her oil was controlled by the Nobel Brothers, and Shell controlled some of the rest." A. Sampson, *The Seven Sisters; the great oil companies and the world they made*, Coronet Books, Sevenoaks, 1985 (seventh impression), p.86. The distribution network of Shell in the world is extensive and after the second world war Shell had failed to find enough crude oil to supply its distribution network. "(...) and in the late 1950s its need for new supplies was far more important than any esoteric commercial principle. In 1960 it became the first major company to accept a host government as a partner when it took out a concession for Kuwait's off-shore areas." C. Tugendhat and A. Hamilton, op.cit., p.155.

[45] Research of Gulbenkian had resulted in good exploration prospects. Only after the San Remo conference, when the Ottoman empire was divided into mandates between France and Great Britain, was the oil exploration initiated seriously.

[46] S.H. Schurr and B.C. Netschert, op.cit., p.86.

[47] H.F. Williamson, R.L. Adreano, A.R. Baum, G. Close, op.cit., p.304.

property rights offsets.[48] In the US the subsoil property rights are defined by the 'Rule of Capture'. The rule "entitles each owner to extract oil or gas by means of wells on his tract (and bottomed under his tract) without liability to an adjacent owner, even though particular molecules of the extracted resource may have originated in the latter's land."[49] This rule generates wasteful production since it results in an extreme number of wells in an oil deposit and an excessive extraction rate.[50] "In earlier years, new discoveries were developed rapidly under the common law 'rule of capture'. The irregular addition of sizable new supplies had a severely disruptive impact on the producing industry, causing unacceptable surface and sub-surface waste, wide price fluctuations, the premature abandonment of reserves and a generally unstable environment not conducive to sound long-term business decisions."[51] The production capacity of oil deposits deteriorated and reduced the total volume of recoverable oil. As a result, supply was rather volatile with a tendency for over-supply. This created a "boom or bust" situation. Evidently, the rule of capture also increased the investment risk, since a company could be forced to sell the oil at a price that would endanger the return on investments. In the absence of government regulation, supply was rather erratic and generated a higher than necessary extraction rate.

Between 1918 and 1924, the Unites States feared an oil shortage for the near future.[52] Advisers to President Wilson estimated in 1919 that 40% of the domestic oil had already been exhausted, due to increased consumption of oil products. Alternative sources of petroleum, like shale-oil, were perceived as economically unattractive and therefore the development of foreign oil fields by American companies, with the support of the American government, was recommended.[53] Increasing domestic consumption combined with (supposedly) dwindling domestic production would establish dependence on foreign oil. In addition to domestic considerations, the American government had international political reasons to stimulate foreign direct investments in petroleum. A future dependence on, mainly, British oil production was undesirable. Therefore, the American government succesfully pressured the British government to open the Iraq Petroleum Company for

[48] W.S. Peirce, op.cit., p.136.

[49] T. Daintith and S.F. Williams, *The legal integration of energy markets*, series Integration through Law, volume 5, Walter de Gruyter, Berlin/New York, 1987, p.3.

[50] The fixed costs of exploration are high compared to relative costs and the cost of exploration is much higher than operating costs in the production stage. In order to recover the investment, producers, under the 'Rule of Capture', tend to favour present production over future production. The price consequently declines as producers continue their rate of production at almost any price. See Chapter 1.

[51] US Senate, *The Petroleum Industry*, part 1, Economists' views, Hearings before the subcommittee on antitrust and monopoly of the committee on the judiciary United States Senate, ninety-first congress, first session, S.Res. 40, 1969, p.528.

[52] US Senate, *The International Petroleum Cartel*, Staff Report to the Federal Trade Commission submitted to the subcommittee on Small Business, United States Senate, August 22, 1952, p.39-41.

[53] S.J. Randall, *United States Foreign Oil Policy; For Profits and Security, 1919-1948*, McGill-Queen's University Press, Kingston/Montreal, 1985, p.14-15.

participation of American oil companies in order to secure supply. Although the American estimates of the lifespan of their domestic production proved to be wrong, as new oil discoveries were made in the 1920s, the expansion of American companies in the international oil industry in the Middle East and Venezuela continued.

Table 2.9 US replacement ratios 1920-1929

year	ratio	year	ratio
1920	2.1	1925	2.3
1921	2.3	1926	1.4
1922	0.6	1927	2.9
1923	1.0	1928	1.6
1924	0.8	1929	3.2

source: H.F. Williamson, R.L. Adreano, A.R. Daum, G. Close, *The American Petroleum Industry; the age of energy 1899-1959*, Northwestern University Press, 1963, p.304

Wasteful production and worry over the replacement ratios in the 1920s initiated a conservationist movement in the US. This depletion policy[54] of the US at first met opposition from the major oil companies which had partly internalised the costs of misallocations through the acquisition of larger concession areas covering a complete deposit and through direct foreign investments. In other countries the companies could obtain larger concession areas, sometimes comprising an entire province or country.

In the second part of the 1920s, the opposition of the companies to the regulation of US production subsided, as competition in the declining market cycle resulted in lower oil prices.[55] Occasional price wars had not only affected the revenues of the companies but also their exploration efforts. The major American companies and Royal Dutch/Shell and BP had secured and started to develop major and promising concession areas in the Middle East. The companies anticipated problems because: "New oil discoveries quickly seek

[54] "The stated objective of proration in the United States has always been conservation, i.e., obtaining the maximum total yield of oil from each producing field. Valid technological considerations indicate that this objective is best achieved by operating each field as a unit in such a manner as to maintain the natural hydrostatic and gas pressure necessary to assure maximum total recovery. Proration for conservation alone involves only the limitation of production in each field to those allowables which will maintain those pressures, regardless of the price at which the oil can be sold." US Senate, *The International Petroleum Cartel*, 1952, Washington D.C., p.213.

[55] "Up until 1928 the history of the international oil industry was a continuous series of fierce struggles for markets and profits, primarily between Standard Oil of New Jersey and Royal Dutch/Shell." (...) "Moreover, it should be noted that as the oil demand boomed not only for automobiles, but for fueling ships, planes, and tanks, the struggle for position in the international oil industry increasingly became intertwined with national rivalries. This put oil smack in the middle of the constant jockeying for military and economic power which characterized relationships among the developed capitalist countries." M. Tanzer, op.cit., p.101-102.

accommodation in the market place. When discoveries are very large, market adjustments are difficult and are made more so by a demand for petroleum products which is largely unresponsive to price changes, save perhaps for very wide price swings and over a very long term. Thus, the burden of market adjustment falls almost entirely upon the supply side."[56] The foreign markets were not yet sufficiently developed to absorb both American oil and oil from the new producing countries. Although consumption of oil was still growing in the US, the major expansion had already taken place and the other markets were slower to develop. Coal remained the major fuel in Europe until well after 1945. Although the European market was expanding, it could not absorb the quantities available from foreign sources. The industry's declining market cycle was international.

The discovery of the large deposits in Texas in the 1920s had dissolved the fear of the US government for immediate domestic shortages, but did not end the vigor of the conservationist movement. When the oil market started to experience a glut, the opposition of the oil companies to implement 'allowables' in the US diminished. Regulation of American oil production could in fact create the necessary room on the market for their foreign oil, without impeding oil prices. This required output control. The government proposals to implement conservation measures, at first opposed, found support from the international oil companies when the market experienced over-supply. The companies wanted to restrain production and price competition. However, the stringent enforcement of American anti-trust law inhibited the companies to conclude cartels or other collusive market behaviour. This inhibitation did not prevent the companies from concluding international agreements.

In 1928, at Achnacarry Castle in Scotland, Exxon, BP and Royal Dutch Shell agreed to regulate the international market.[57] They assented to regulate their international production to the development of demand, and in accordance with their market shares ('As Is' Agreement). They also decided to price oil from a single basing point (Gulf Plus system). The proposed measures to conserve oil in the US agreed with the companies' efforts to regulate the international market. The 'As Is' Agreement required control of outside competition to maintain the market share quotas and the oil price level. The recommendations of the American Petroleum Institute's Committee on World Production and Consumption of Petroleum and its Products in 1929 covered this loophole in the 'As Is' Agreement.

[56] US Senate, *The Petroleum Industry*, Washington D.C., 1969, p.528.

[57] The presidents of Shell, BP and Exxon decided to restrain price competition. The 'As Is' Agreement (backed up by more oil companies) wanted to curb excessive competition. Market shares were to be stabilized at their 1928 level and increased production by the participating companies would only serve increased demand. The basis of the Agreement, which was never fully put into practice, was protection of the American price level for oil. The Gulf Plus system fixed oil prices, whatever the production costs, at the level of American oil prices in the Gulf of Mexico. Transportation costs were based on the transportation costs from the Gulf of Mexico to the port of entrance, whether the oil was shipped from the Middle East, Venezuela or the Gulf of Mexico. In this way the interests of the oil industry in the United States, still the most important producer as well as consumer, was freed from competition from low cost oil production areas.

Production in 1929 and in subsequent years should be maintained at the 1928 level.[58] The US Senate report on the International Petroleum Cartel continues: "There is a striking similarity between this recommendation and the "as is" quota provisions of the Achnacarry Agreement, for under the latter 1928 was also to be the base year for subsequent international quotas set up to implement its principles. One dissimilarity, however, is to be noted; namely, that the conservation recommendations for the United States would permit no increased production over that of 1928, whereas the Achnacarry Agreement contemplated increases in production to cover world consumption. If United States production could be held rigidly to the 1928 level there would, of course, be no "surplus" production available for export to trouble the cartel. Thus the proposed American conservation control plan would have had the effect of plugging the loophole which, for legal reasons, had been left in the cartel agreement by the deliberate exclusion of the domestic market of the United States and imports into the United States. The conservation plan was promptly approved by the American Petroleum Institute on March 27, 1929."[59]

The conservation policy was adopted and operated under an interstate compact. Rules were established for 'well-spacing' and 'allowables' in the late 1920s.[60] The output restrictions in the US had been adopted to curb drilling and production from each producing field to obtain a maximum yield per field, but some states extented these restrictions to maintain prices. The Connally 'Hot Oil Act' of 1935, which prohibited interstate shipments of oil in excess of state regulations, facilitated higher oil prices.[61] In the second part of the 1930s, the market situation improved and prices increased. However, the conservation policy in the US, and the arrangements between the large oil companies at the end of the 1920s and beginning of the 1930s to control exploration and production in the Persian/Arabian Gulf (Red Line Agreement), to stabilize oil prices (Gulf Plus System) and to protect their market shares ('As Is' Agreement and the subsequent market charters[62]) laid the foundation for the post-war operations of the international oil industry.

7. Conclusion

The industrialisation process in the US in the second half of the nineteenth century was accompanied by rapid growth of mineral fuel production and consumption. Coal quickly gained importance as the main fuel in the economy. The oil industry, also expanding, was

[58] US Senate, *The International Petroleum Cartel*, 1952, Washington D.C., p.212.

[59] Ibid., p.212.

[60] Well-spacing requirements prohibit drilling of wells at greater than a specified density. Allowables constitute ceilings on the amount of production from each well. T. Daintith and S.F. Williams, op.cit., p.87.

[61] W.S. Peirce, op.cit., p.141.

[62] "Member companies were allowed to expand only in proportion to the growth of the total market. This rule could only be violated in the event of an increase in market share being won at the expense of companies not participating in the agreement." A.Roncaglia, op.cit., p.55.

mainly processing oil into lamp oil. Between 1860 and 1880, the industry experienced a period of impressive expansion of consumption and production. Prices were very volatile in this period, within a downward trend. The market was competitive and deconcentrated. From 1880 onwards, growth of production and consumption stagnated. The US lamp oil market had matured. In these circumstances, price competition became an increasingly inefficient strategy to drive competitors out of the market as the producers lacked the option of cost decreases, and lamp oil already had a low elasticity of demand. Price competition only squeezed the profits. In order to gain market share, Rockefeller embarked on take-overs in the late 1870s, and concluded a cooperative agreement with a rail company to tranport crude against fixed prices. This strategy gave the Standard Oil Trust virtual monopoly power in the industry. Between 1860 and 1900, the oil industry had completed a full growth cycle.

At the turn of the century, major new oil discoveries and technological advances in refining and the engineering industries changed the market conditions, and a new growth cycle began. The new entrants seized the opportunity to gain a market share in the new fuel product markets. The use of oil as a fuel brought about a new period of expansion for the American oil industry, which lasted until 1930. In this period, coal was progressively substituted by oil. In the 1920s, the primary energy market began to show signs of maturity, which also affected the oil market developments.

In this decade, US energy consumption per unit of GNP began to decrease, while oil and gas consumption continued to increase. The composition of primary energy consumption changed in favour of oil and gas, and consumption of coal per unit of GNP decreased rapidly. As a result of the World War I, oil prices were relatively high until the mid-1920s. This stimulated US exploration. The subsequent increase in production was not accompanied by a similar growth of consumption. The market changed from a buyers' marker into a sellers' market. The economic recession, and shortly afterwards the deep depression in the late 1920s and early 1930s, caused stagnation in the oil industry.

The American oil market was not only the laıgest in the world, but it was also its principal exporter. This situation changed drastically after the World War II. From 1940 onwards, consumption, propelled by the expansion of light transport fuels consumption, grew at a faster pace than domestic oil production. In 1947, the US became a net-importer of oil. Government policies to stimulate domestic production could not prevent the rise in the share of imports in consumption. In 1990, domestic production covered only 54.6% of consumption. Production in Alaska has already peaked, and domestic production in the rest of the States is falling. Despite substitution and conservation efforts, the rate of decline in production was not matched by a similar decline in consumption. Although the expansion of consumption has not resumed its former upward rate, but has grown at a lower level, the call of the US on the world oil market is steadily increasing again.

Chapter Three

International Market Developments

1. Introduction

The history of the oil industry has been dominated by the early development of the US market. From 1860 to 1953, the US share in world oil production was more than 50%, with the exception of the period 1897 to 1903[1]. The rapid expansion of the US market must be seen within the context of the availability of oil, legislation, the prevailing business environment and the technological advances of the US engineering industries at the beginning of the twentieth century, which stimulated demand for liquid rather than solid fuels.[2] Although the oil industry was also developed in other regions of the world, the expansion of the industry was limited by the absence of nearby consumer markets, the competition of domestic produced solid fuels, and transportation constraints.

The development of the (low cost) oil resources in the Arabian/Persian Gulf area in the 1930s by a group of European and American oil companies intensified the internationalization of the oil industry, and generated a structural change of the world oil market. From the 1940s onwards, the expansion of world oil production was remarkable (See Figure 3.1). This expansion was arrested in the 1970s. In the early 1970s the market for oil matured, and subsequently stagnated in the 1980s, ending that particular growth cycle.

World oil consumption experienced a similar period of expansion, maturity and stagnation. The expansion of consumption was a result of expansion of consumer markets, of substitution of coal for oil, and a relocation of the processing sector, which created economies of scale both in processing and in transport. From 1945 onwards, the oil market was progressively operating as a global market, instead of the earlier division into regional markets. The development of oil production in new regions and linking these together into the international oil market created new market conditions, which in its turn generated a new market structure, thus beginning a new growth cycle.

[1] *Petroleum Facts & Figures*, 1971 edition, American Petroleum Institute, p.556 and 557. The share of the US in this period was fluctuating between 40 and 50%.

[2] See Chapter two.

2. World production and consumption after 1945

The post-war expansion of the oil market was impressive. Although the annual average growth rates on the international oil market did not match the rates in the earlier period of expansion on the American market, the wider penetration of oil in the international economy, and subsequently the larger volume of the market, gave the oil industry a fresh impetus. This new phase of expansion attracted new entrants into the market, challenging the market positions of the traditional companies. Also, the wider penetration of oil in the international economy generated market intervention by governments of both producer and consumer countries, when they tried to capture a larger share of the economic rents and profits. These developments were also reflected in the market structures.

World oil production increased from 9,750 thousand b/d in 1949 to 62,195 thousand b/d in 1988, an increase of 537.9% (See Table 3.1).[3] World oil production peaked in 1979, when 65,775 thousand b/d were produced. The magnitude of expansion of oil production did, however, fluctuate in the period 1949-1988 and again distinct periods of expansion, stagnation and contraction of production can be distinguished in this third growth cycle.

Between 1950 and 1973, oil production expanded rapidly (See Figure 3.1). Oil production increased in this period from 10,910 thousand b/d to 57,840 thousand b/d.[4] Between 1970 and 1973, growth of production slowed down, from average annual increases exceeding 10% in the previous two decades to an average annual increase of 7%.[5]

The impact of the oil price rise in 1973/1974 on consumption, and the take-over by the governments of the producing countries of output administration, manifested itself in diminishing growth rates of production.[6] Although production reached a historical peak in 1979, production grew only 8.4% between 1973 and 1980, an average annual increase of 1.2%. Compared to the significant growth in the two previous decades, growth of oil

[3] Calculations in this chapter are based on various issues of the *British Petroleum Statistical Review of World Energy (BPSR)*.

[4] The average annual growth of oil production was 10.2% between 1950 and 1960, and 11.7% between 1960 and 1970.

[5] In Chapter 2, the relationship between economic growth and oil consumption was discussed. Up to 1973, the U.S. oil/GNP ratio was increasing. Diminishing growth of GNP will, therefore, also affect oil consumption and production. Growth of real GNP in the United States decreased substantially from 4.2% in 1968 to 2.4% in 1969, -0.3 in 1970 and 2.8% in 1971. The inflationary economic policy of the United States at the end of the 1960s and the subsequent emergence of a fundamental balance of payments deficit had put increased pressure on the value of the dollar. The economic problems of the United States were passed on to other economies through the Bretton Woods exchange rate system. In this period it became obvious that the exchange rate of the dollar, also important for the oil exporting countries, was grossly over-valued. The resultant confidence crisis in the exchange rate system led to a major monetary crisis. In 1971, the inconvertibility of the dollar in gold became a fact and the dollar devaluated around 7.5% in relation to the OECD countries' currencies. The period of economic uncertainty had also affected the performance of other economies. At the end of the 1960s, growth of real GNP/GDP had slowed down in the OECD countries (1968: 5.4%; 1969: 4.7%; 1970: 2.7%; 1971: 3.5%; 1972: 5.1%).

[6] See Chapter 1, section 4.

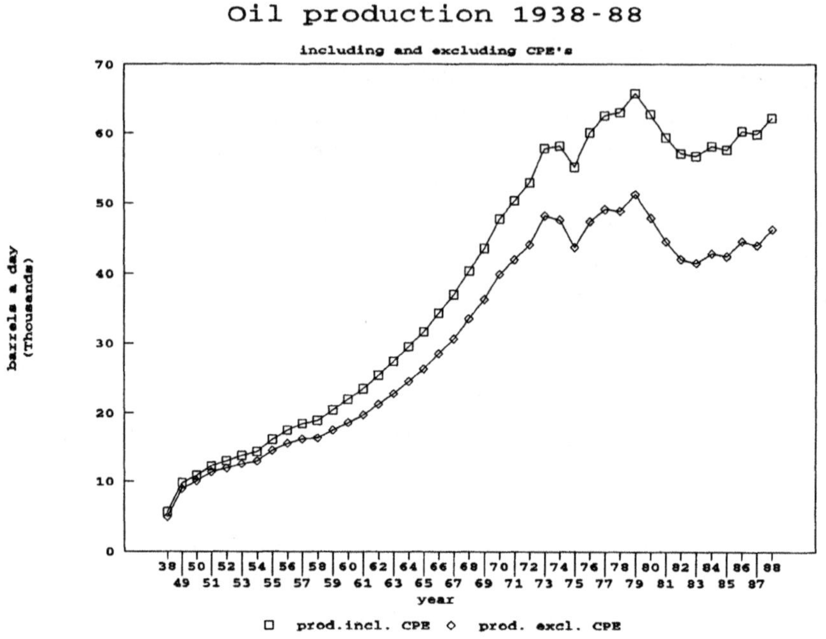

Figure 3.1

production turned out to be considerably less between 1970 and 1980 (31%).

From 1980 to 1986 production actually declined 4.2% (average annual decrease of 0.7%), reflecting the impact of the second price rise (See Figure 2.1) on consumption (See Figure 3.2) and the ensuing output restrictions by OPEC to maintain the oil prices. Due to the collapse in oil prices in 1986 and the subsequent increase in consumption, production increased 3.1% again between 1986 and 1988.

Oil consumption can be divided, like production, into periods of expansion, stagnation and contraction of consumption. Between 1949 and 1988, oil consumption increased from 9,880 thousand b/d to 63,245 thousand b/d, an increase of 540.1% (See Figure 3.2); a slightly higher growth rate than oil production in this period.[7] Compared to growth in production between 1970-1973, consumption grew at a noticeably higher rate (32.5% compared to 21%), showing that the market had turned into a sellers' market.

Following the oil price increases in 1973/1974, oil consumption contracted, after the years of slowly stagnating growth in the late 1960s and the short boost in demand in the early

[7] Oil consumption expanded rapidly from 1950 to 1973, 95.6% between 1950 and 1960, 115% in the period 1960-1970 and 32.5% between 1970 and 1973.

1970s. The turbulence on the oil market in the early 1970s was part of a general crisis in the international economy. This crisis had already caused economic growth to stagnate, had led to a major adjustment of exchange rates in 1971, and had resulted in the demise of the Bretton Woods system in March 1973. Real growth of GNP/GDP in the OECD (Organization for Economic Co-operation and Development) countries diminished after the tumultuous events in 1973. In 1973 real growth (in percentage change from the previous year) in the OECD countries was still 5.8%, but declined to 0.6% in 1974 and to -0.5% in 1975.[8] World oil consumption declined 1.1% in 1974/73 and 1.2% in 1975/74, while NCW (World excluding Centrally Planned Economies; i.e. Non-Communist World) oil consumption decreased more (3.3% in 1974/73 and 2.6% in 1975/74). NCW consumption (See Figure 3.2) reacted with more intensity because the higher oil prices were immediately passed on to the consumers[9] and the economic recession, already imminent in the beginning of the 1970s, was aggravated when oil prices increased suddenly.

[8] *OECD Economic Outlook*, 43, June 1988, Table R 1, p. 170.

[9] In the Centrally Planned Economies, the oil price increase of 1973/1974 was not immediately passed on to the consumers. Among the CMEA countries, prices were determined for a five year period and based on the average world market prices of the previous five years. P. Boot, Economische stagnatie in Oost-Europa, in: *Internationale Spectator*, Oktober 1983, xxxvii-10, p. 620.

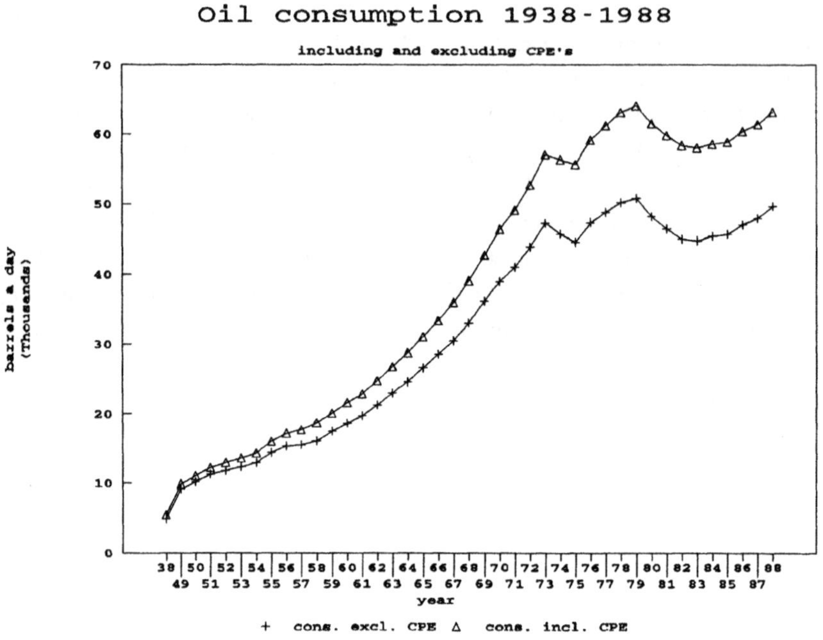

Figure 3.2

In 1976, real growth recovered to 4.7% and remained firm until 1979 (3.2%). Due to inflation, real oil prices had declined somewhat. As a result, oil consumption recovered again. However, the 1970s saw oil consumption grow at substantially lower rates than in the previous decade.[10] Due to the price increase of 1979/1980 and a deceleration of real growth of GNP/GDP in the major consuming countries, oil consumption contracted from 1980 to 1983, from 61,545 thousand b/d to 58,105 thousand b/d.[11]

Recovery of real economic growth from 1984 onwards and weakening nominal and real oil prices (See Figure 2.1) were reflected in an increase of oil consumption in the period 1984-1988 (from 58,630 thousand b/d to 63,245).

World oil production exceeded consumption from 1951 to 1980, except in 1975 and 1978. Between 1967 and 1971, oil production was in a substantial excess over consumption. In 1972, this surplus had diminished to a mere 200.000 b/d, and the market had become a sellers' market. However, in 1973 the surplus of production over consumption had again

[10] Consumption increased only 32.5% between 1970 and 1980.

[11] Real GNP/GDP diminished to 1.1% in 1980, 1.6 in 1981, -0.5 in 1982 and recovered from 1984 (4.9%) onwards. In 1987, real growth of GNP/GDP in the OECD area was 3.1%. *OECD Economic Outlook*, 43, June 1988, Table R 1, p. 170.

increased, to 795.000 b/d. In 1974, 1977, 1979 and 1980, production exceeded consumption more than 1 million b/d. From 1980 onwards world consumption has exceeded world production by, on average, more than a million barrels a day.

3. Development of production and consumption excluding CPE's

NCW oil production increased 417% between 1949 and 1988, significantly less than world oil production. The major expansion of NCW oil production also took place in the 1950s and 1960s.[12] In the three years preceding the oil price increase of 1973/1974, growth of oil production was slowing down, and was slightly negative (-0.5%) between 1973 and 1980. The difference in growth between world oil production and NCW oil production (a 8.4% increase of world oil production compared to a -0.5 % decrease of production excluding CPE's) was mainly accounted for by a decline of OPEC and North American production and a 6.7% increase in production by the CPE's.[13] From 1980 to 1985 NCW production decreased 11.4%, compared to a decrease in world production of 8.2%. In this period, the production of North America, Latin America, Western Europe, the CPE's[14] and Asia/Australia increased, while the production of Africa and the Middle East declined (See Figure 3.16). OPEC production declined from 27,445 thousand b/d in 1980 to 17,215 thousand b/d in 1985 (-37.3%). However, the 10,230 thousand b/d decline was mainly accounted for by the Middle East OPEC members, who bore the brunt of the output restriction policy. Their production declined 7,980 thousand b/d (-43.7%). In December 1985, the OPEC countries, under pressure from its Middle East members, abandoned this policy[15], and tried to increase the OPEC market share by increasing production and leaving oil prices to the market. As a result, oil prices (in nominal and real terms) declined sharply (See Figure 2.1), and subsequently NCW oil production recovered somewhat in the period 1986-1988.

Before 1970, NCW oil consumption grew at similar rates to production.[16] Also between

[12] Between 1950 and 1960, production increased 85.4% and 114.1% between 1960 and 1970.

[13] OPEC production declined from 30,988 b/d in 1973 to 27,445 b/d in 1980, a decline of 3,543 b/d. CPE production increased 5,165 b/d in the same period (9,645 b/d in 1973 to 14,810 b/d in 1980. See Figure 3.16 and 3.18.

[14] "On the domestic side, oil is the swing fuel with flexibility of delivery, though Soviet planners have no flexibility of supply through further expansion of production, which is subject to severe physical and economic constraints. The flexibility of supply of Soviet oil exports thus reflects flexibility in the allocation of delivery; and the surge of oil exports to the hard currency area in the early 1980s, while production remained stagnant, is largely explained by the diversion of oil deliveries to the West at the expense of the CMEA countries and domestic consumers." M. Chadwick, D. Long, M. Nissanke, *Soviet Oil Exports; Trade Adjustments, Refining Constraints and Market Behaviour*, Oxford Institute for Energy Studies/Oxford University Press, 1987, p.85.

[15] See Chapter 5.

[16] Between 1950 and 1960, consumption increased 84.2%, 109% between 1960 and 1970 and 21.4% between 1970 and 1973.

1970 and 1973, growth of NCW oil production and consumption was quite balanced (20.6% compared to 21.4%). This was in contrast to world oil production and consumption, which was less balanced (21% compared to 32.5%). This was due to the fact that oil production in the CPE's increased at a slower growth rate than consumption.[17] After 1973, NCW oil production has been much more volatile than world production. The role of OPEC as 'swing producer' in the 1970s and the beginning of the 1980s, and the change in policy in 1985 to defend their market share caused this volatility of production.

NCW oil demand was quite responsive to the oil price increases of 1973/1974 and 1979/1980 and to the general economic downturn. The price increases of 1973/1974 and 1979/1980 did depress demand considerably, necessitating first a deceleration of production growth and then actual output cuts in order to maintain the oil price levels.[18] However, the reaction to the second oil price increase, especially in the OECD countries, turned out to be more structural than to the first raise. Consumption excluding CPE's, after a short recuperation in the 1970s, declined 4.3% in the period 1980-1985. The second oil price rise had forced the governments in the major OECD countries to implement austerity policies in order to reduce government budget deficits and to bring down inflation[19]. The consequent contraction of economic growth, and substitution away from oil resulted in downward adjusted oil demand. Consumption in the OECD countries decreased from 1980 to 1983[20], increased in 1984 (2.1%), experienced negative growth in 1985 (-1.0%), but increased again from 1986 onwards.[21] The decline in oil intensity[22] of the OECD countries, which had already decreased from 46 in 1973 to 41 in 1979, accelerated after the second oil price increase and decreased from 37 in 1980, to 34 in 1981 and to 32 in 1982.

The combined effect of collapsing oil prices and a depreciating dollar in 1985 was a positive incentive to oil consumption growth. From 1986 to 1988, consumption increased 4.6% without really turning the trend of decreasing oil intensities in the OECD countries.

[17] In the period 1970-1973, oil production in the CPE's increased 22.9% (from 7,845 thousand b/d to 9,645 thousand b/d) and oil consumption 28.4% (from 6,805 thousand b/d to 8,735 thousand b/d).

[18] OPEC's production cuts are reflected more prominently in the figures of the world excluding CPE's than in total world production. The production cuts by OPEC were partly substituted by increased production by the USSR.

[19] Consumer prices increased at an average annual rate of 6.6% in the period 1966-1975 and 8.5% in the period 1975-1984 for the OECD countries. In Japan consumer prices increased at an average annual rate in these periods of 8.8% and 4.9% respectively, in the United States 5.8% and 7.6% respectively and in the European OECD countries 7.1% and 10.0% respectively. *OECD Economic Outlook*, 43, June 1988, table 22, p. 49.

[20] OECD consumption decreased 7.6% in 1980, 5.5% in 1981, 5.3% in 1982 and 1.7% in 1983. *BPSR 1989*, p.9-10.

[21] OECD consumption increased 3.1% in 1986, 1.1% in 1987 and 2.9% in 1988. *BPSR 1989*, p. 9-10.

[22] Oil consumption in million tonnes oil equivalent per US$ 100,000 of GDP at 1975 prices and exchange rates. G. Koopmann, K. Matthies and B. Reszat, *Oil and the International Economy, lessons from two oil price shocks*, Transaction Publishers, New Brunswick/Oxford, 1989, p.36.

Only in 1986, a minor increase in the OECD oil intensity was recorded.[23] Demand in the developing countries, especially in the fast growing Asian countries, is still strongly related to economic growth. Buoyant economic growth combined with lower oil prices in the second part of the 1980s prompted increased oil demand.

The development of world oil consumption deviates from the development of NCW oil consumption in the period 1973-1980. World consumption (7.8%) is no longer paralleled by growth of NCW oil consumption (1.9%). The reaction of the centrally planned economies to the raised oil prices in 1973/1974 was substantively slower than the market economies. The intra-CMEA pricing system has certainly been an influencing factor. In 1975, the Bucharest pricing formula was revised, which determined the price of Soviet oil contracted by the East European countries under the Five-Year Plans at a five-year moving average of world oil prices prior to the year in question.[24] The centrally planned East European importing countries enjoyed, under this formula, oil prices lower than the world market prices until 1980. During this period, these countries were engaged in a process of modernizing their industry based on imported technology from OECD countries. This technology was, however, designed to conform to patterns of energy consumption in the OECD countries, which largely consisted of liquid fuels, thus stimulating increased oil consumption in a period when oil became more expensive.[25] The economic effects of this changing consumption pattern in the early 1970s became clear when, in 1980, oil price increases were passed on to these countries more quickly.[26]

The response to the second oil price increase and general economic recession bears more similarity as consumption decreased -2.6% (world) and -3.8% (NCW) respectively, in the period 1980-1986. The second oil price increase resulted in a 25-28% price increase in 1979 and a 21-27% price increase in 1980 in intra-CMEA trade. On top of that, the Soviet Union increasingly diverted exports destined for East European countries to the world market. These countries were either forced to buy oil on the world market or to clamp down

[23] The OECD oil consumption/GNP ratio (1980=100) was 77.6 in 1985, 77.8 in 1986 and 76.4 in 1987. *OECD Economic Outlook*, 43, June 1988, table 71, p. 157.

[24] G.W. Hoffman, *The European Energy Challenge; East and West*, Duke Press Policy Studies, Durham, N.C., 1985, p. xxiv, note 3.

[25] Ibid., p. 17.

[26] "Nevertheless, the price increases for Soviet oil in the past three years have been greater than could have been foreseen from the principle of a moving five-year range. The mode of calculation has been changed since 1980; since then prices seem to have been calculated according to average world market prices for the previous three years, rather than five. In the past three years this has led to distinctly steeper price rises. (...) Thus the following hypothesis may be proposed: The change in the price mechanism forced the smaller CMEA countries to cut back their oil imports from the USSR. If the three-year average is maintained, clearing prices will nearly be 10 per cent higher than the world market price by 1983. Price declines are to be expected for 1985 and 1986." J. Bethkenhagen, The impact of energy on East-West Trade: retrospects and prospects, in: C.T. Saunders (ed.), *East-West Trade and Finance in the World Economy, a new look for the 1980s*, Macmillan, London, 1985, p. 186/187.

consumption immediately.[27] The response of demand of the CMEA countries to the price collapse of 1986 was comparable to the reaction of the NCW countries.

In 1988/1987 world oil consumption increased 3.1%, compared to an increase in total primary energy consumption of 3.7%. In the NCW economies oil consumption increased 3.9%, compared to 0.2% in the centrally planned economies.[28] This growth of demand confirmed the fact that oil prices had been over-valued before 1986 and had suppressed demand for oil. It should be noted, however, that demand is to some extent inflexible, unable to respond quickly to price changes because of investment commitments to other fuels.

4. The balance between supply and demand

The imbalance between NCW oil production and consumption (See Table 3.1) has, since 1979, widened. From 1951 to 1959, oil production exceeded oil consumption. In 1957, the surplus was substantial, 750.000 b/d. Between 1960 and 1967, in the wake of the 'isolation' of the US market from the world market, production experienced alternating small deficits and surpluses over consumption, while world oil production continued to be in surplus over world consumption.[29] These deficits in the NCW market can be attributed to the production policy of the Middle East consortia, who had to adjust output to compensate for the near-closure of the American market for Middle East oil, and for the increase of exports from the Soviet Union. The constraints on production growth of the IPC and KPC[30] in the 1960s, for example, must be seen as an attempt by the large oil companies to maintain oil prices. However, from the mid-1960s onwards it is clear that this attempt failed due to increased competition from producers outside the Middle East consortium companies. The surplus of production over consumption started to rise, and discounts on the posted oil price increased. Also real oil prices declined. In 1970 and 1971, NCW production was in a substantial surplus over consumption of 920.000 b/d and was surpassed by an unprecedented surplus of 1.8 million b/d in 1974. Between 1975 and 1977, only small deviations, compared to the 1974 surplus, in the balance occurred. In 1978, a deficit of production over consumption of 1.4 million b/d upset the balance.

From 1977 to 1978, OPEC production decreased by exactly 1.4 million b/d, a 4.4% decrease. Saudi Arabia, with its vast balance of payments surplus, was responsible for 66%

[27] From 1980 to 1985, apparent consumption of oil in the 6 Eastern European CMEA countries declined from 1930.4 thousand b/d to 1742.6 thousand b/d, while imports slightly decreased. PlanEcon, *Soviet and East European Energy Databank*, Volume two, March 1986, Washington, D.C., p. x-1.

[28] Primary energy consumption increased, 4.1% in the market economies and 3% in the CPE's. In the OECD countries oil consumption increased 2.9% in 1988/1987, while primary energy consumption increased 3.4%.

[29] In 1960 and 1961, production was just in deficit over consumption, a mere 15.000 b/d. In 1962 there was a surplus of 50.000 b/d, in 1963 a deficit of 170.000 b/d, in 1964 a deficit of 55.000 b/d, in 1965 a more substantial deficit of 360.000 b/d, in 1966 a small deficit of 55.000 b/d and in 1967 production was in surplus to consumption again (170.000 b/d).

[30] See Chapter 4.

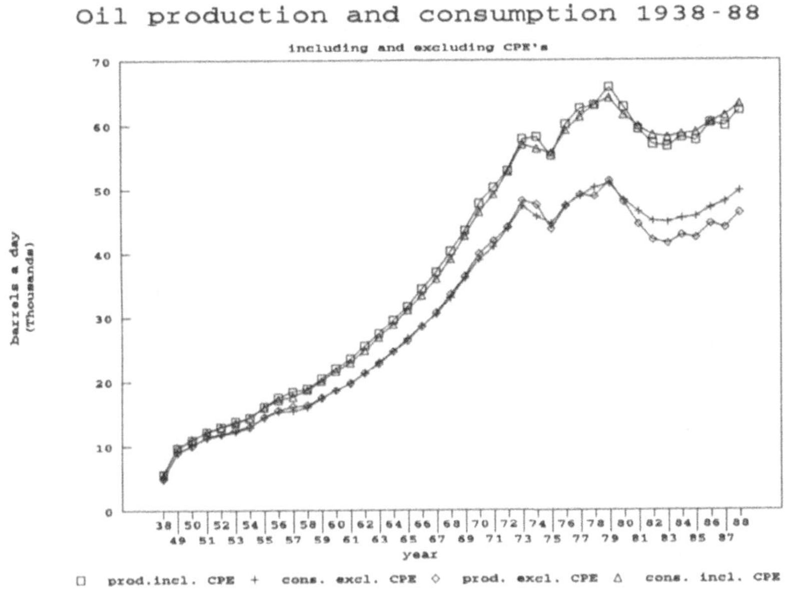

Figure 3.3

of the decrease in OPEC production. The balance of world production and consumption was also experiencing a small deficit, and the market was tight. The reduction in OPEC production had to support the pricing goals. Between 1974 and 1978, real oil prices had declined, despite nominal oil price increases.

The second round of oil price increases in 1979 and 1980 cannot easily be explained from the balance between world production and consumption. In 1979, despite production problems in Iran, world production was 1.6 million b/d in surplus over consumption and in 1980 still 1.2 million b/d. In 1979, NCW production was 395 thousand b/d in surplus over NCW consumption, but 320 thousand b/d in deficit in 1980.

In 1980, the former balance between supply and demand in the NCW market was disrupted (See Figure 3.3). In 1981, the deficit was 2 million b/d, increased to 3 million b/d in 1982, to 3.3 million b/d in 1983, diminished to a deficit of 2.5 million b/d in 1984, increased again to 3.3 million b/d in 1985 and to 4 million b/d in 1987. In 1988, the deficit of production over consumption had declined a little to 3.4 million b/d. In this period, stocks in the OECD countries increased above the 90-day OECD requirements, creating a huge

Table 3.1 World oil production and consumption 1938-1989; in thousand barrels per day

year	production including CPE	production excluding CPE	consumption including CPE	consumption excluding CPE	p>c including CPE	p>c excluding CPE
1938	5,595	4,875	5,440	4,890	155	-15
1949	9,750	8,960	9,880	9,080	-130	-120
1950	10,910	10,070	11,040	10,140	-130	-70
1951	12,260	11,380	12,210	11,210	50	170
1952	13,000	11,890	12,900	11,720	100	170
1953	13,800	12,550	13,570	12,260	230	290
1954	14,420	13,010	14,350	12,960	70	50
1955	16,180	14,510	16,030	14,410	150	100
1956	17,530	15,580	17,220	15,340	310	240
1957	18,495	16,230	17,660	15,480	835	750
1958	18,945	16,360	18,740	16,070	205	290
1959	20,485	17,535	20,100	17,430	385	105
1960	22,020	18,665	21,600	18,680	420	-15
1961	23,485	19,745	22,910	19,760	575	-15
1962	25,490	21,310	24,780	21,260	710	50
1963	27,405	22,810	26,860	22,980	545	-170
1964	29,525	24,555	28,780	24,610	745	-55
1965	31,685	26,290	31,110	26,650	575	-360
1966	34,385	28,515	33,410	28,570	975	-55
1967	37,035	30,675	36,020	30,505	1,015	170
1968	40,345	33,545	39,105	33,090	1,240	455
1969	43,605	36,375	42,800	36,185	805	190
1970	47,800	39,955	46,440	39,035	1,360	920
1971	50,325	41,980	49,155	41,060	1,170	920
1972	52,965	44,100	52,765	43,925	200	175
1973	57,840	48,195	57,045	47,375	795	820
1974	58,165	47,610	56,395	45,800	1,770	1810
1975	55,205	43,760	55,725	44,620	-520	-860
1976	60,080	47,405	59,165	47,425	915	-20
1977	62,560	49,135	61,230	48,880	1,330	255
1978	63,050	48,850	63,145	50,285	-95	-1435
1979	65,775	51,280	64,115	50,885	1,660	395
1980	62,745	47,935	61,545	48,255	1,200	-320
1981	59,375	44,525	59,850	46,535	-475	-2010
1982	57,060	42,075	58,425	45,075	-1,365	-3000
1983	56,685	41,540	58,105	44,875	-1,420	-3335
1984	58,105	42,870	58,630	45,495	-525	-2625
1985	57,589	42,465	58,900	45,800	-1,305	-3335
1986	60,305	44,645	60,455	47,145	-105	-2500
1987	59,895	44,025	61,455	48,075	-1,560	-4050
1988	62,195	46,320	63,245	49,765	-1,050	-3445
1989	63,895	48,235	64,805	51,855	-910	-3620
1990	64,875	50,000	64,680	52,640	195	-2640

source: BPSR various issues, own calculations

volume hanging over the market.[31] In 1983 on-land oil stocks were a record 102 days of forward supply. In 1984 the stocks were drawn to 91 days, and in 1988 stocks had increased again to 99 days of forward supply.[32] The main cause for the imbalance in the NCW market was the growth of CPE exports. CPE exports, predominantly the Soviet Union, infringed on the market share of NCW producers.[33] In 1980, CPE exports had a share of 6.3% in total trade and this share increased to 13.9% in 1988.

From 1981, the world market also experienced a deficit of production over consumption. However, the size of these deficits was much lower than in the NCW market, and refinery gains and some de-stocking should be taken into account.[34] Given the level of the deficits, the world market was fairly balanced.

The relatively small surpluses and deficits in NCW production over consumption between 1951 and 1965 and the continued surplus of world production over consumption in the same period affirm that the large international oil companies effectively regulated supply in this period. The growing surpluses in the world market as well as the NCW market after 1965 were an indication of the weakening regulatory power of the large oil companies. Production from the independent oil companies and increasing exports from the Soviet Union were no longer compensated.[35]

From 1973 onwards, the balance between production and consumption is dissolved. Large surpluses in the 1970s and deficits in the 1980s characterized the oil market both including and excluding CPEs. Indeed, OPEC, as the proclaimed oil market regulator, has been unable to match demand and supply in the international market. The initial decline of consumption after 1973 and the recovery of demand after 1975 required substantial output adjustments. These adjustments were for a large part made by Saudi Arabia and Iran. The difficulty in matching supply and demand at the new oil price levels was enhanced by increased production by the CPE's and non-OPEC producers (See Figure 3.16).

OPEC production was fairly stable between 1976 and 1979, but decreased sharply thereafter. In 1980, the majority of the OPEC member states decided to defend new oil price levels and embarked on a restrained production policy from 1982 onwards. This policy

[31] Stocks increased from 95 days in 1985 to 99.1 in 1986 and 98.3 in 1987. *OECD Economic Outlook*, 43, June 1988, p. 157.

[32] T. Morgan, Hurdles along the path to $18 oil, in: *OPEC Bulletin*, March 1989, graph 4, p.7.

[33] CPE oil exports increased from 2,005 thousand b/d in 1980 to 3,802 thousand b/d in 1988. Between 1980 and 1984 exports grew 42.9% and 32.7% between 1984 and 1988. In rate of change in 1988/1987 alone was 30.7%. *BPSR 1989*.

[34] The world market can not sustain a deficit of production over consumption for very long. The imbalance can be explained by stocking/de-stocking, refinery gains, unaccounted for stocks and the inability to measure demand and supply precisely.

[35] Growth of Soviet oil exports to the OECD countries increased rapidly in the period 1965-1968 and was followed by a period of more modest export growth. Between 1962 and 1970 the volume of Soviet oil exports to the OECD increased from 14.3 million tonnes to 38.3 million tonnes. M. Chadwick, D. Long and M. Nissanke, op.cit., p.39.

collapsed when the price of oil started to decline, and the member states increasingly disagreed about the desired production policy. The widening deviations in economic development and income requirements inhibited effective supply regulation by the Organization.[36] In 1985, the oil price was left to the market.

OPEC production was very volatile between 1985 and 1989, ranging from 14 million b/d in 1985 to almost 23 million b/d in 1987, and dropping below 20 million b/d at the end of 1988. In 1989, OPEC production recovered and has been considerably above 20 million b/d since. Non-OPEC production (excl. CPEs) was more stable in this period, fluctuating slightly between 22 and 23 million b/d.[37]

5. Development of consumption in the main consumer markets

The expansion of oil consumption in the main consumer markets had taken place in intervals. In the US the rapid expansion had occurred before 1940, and the oil market experienced substantially lower annual average growth rates after 1945.[38] The ratio oil/GNP continued to increase until 1973 (See Table 2.3) but declined afterwards. In Western Europe, the expansion of oil consumption came with the economic reconstruction after the war, and in just two decades oil consumption in Western Europe almost caught up with American consumption (See Figure 3.4).[39] Growth of West European consumption was impressive, and much faster than American growth in the same period.[40] From the 1960s onwards, the Japanese market also expanded at a fast rate; 578% from 1960 and 1970 and 36.5% from 1970 to 1973. The year 1973 marked the end of the expansionary period of consumption in these three major consumer markets. Demand started to stagnate and actually declined in the period 1980-1984. From 1985 onwards, demand recovered, but annual growth rates compared poorly with the buoyant pre-1973 levels.

The industrialization efforts of the developing countries have resulted in a continued stronger demand for oil (See Figure 3.7) throughout the 1970s and 1980s. Oil price increases seem to have only affected the pace of growth, but no signs of stagnation are apparent.

The dominance of the American market before 1945 was increasingly challenged by the expansion of other markets. In 1938, US oil consumption accounted for 57.4% of total world consumption. Between 1950 and 1960, this share had decreased from 58.9% to 44.7%, and

[36] See Chapter 5.

[37] T. Morgan, op.cit., p.6, Graph 2.

[38] See Table 2.6.

[39] R. Lefeber and J.G. van der Linde, op.cit., p 452-458.

[40] Growth of West European consumption was in 1950-1960: 169.2%; 1960-1970: 257.4%; and 1970-1973: 19.1%.

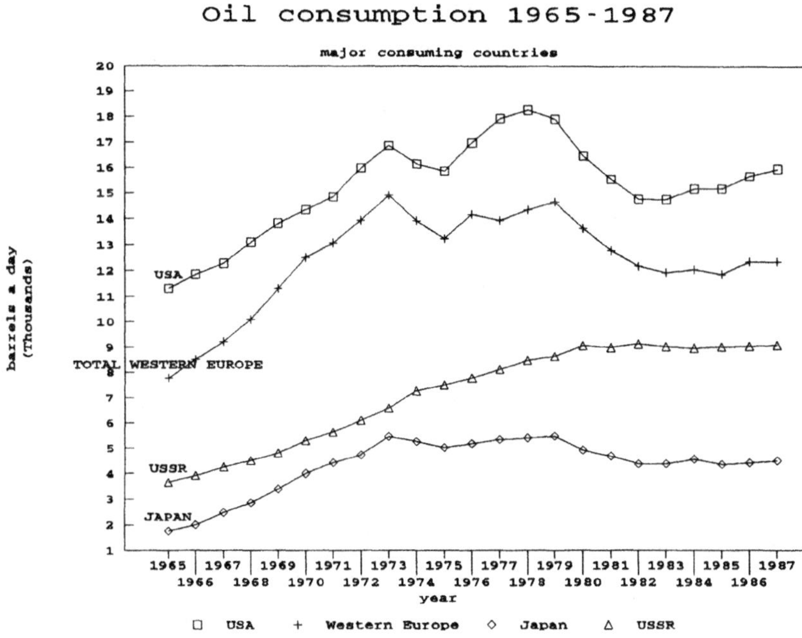

Figure 3.4

it continued to decline. In 1988, this share had dropped to 26% of world consumption.[41]

The relative maturity of the US market was confirmed by lower growth rates than NCW oil consumption. Also in the US, the combination of the oil price increase and the economic recession in 1974 and 1975 temporarily depressed demand (See Figure 3.5). In 1976 demand recovered and consumption reached a peak level in 1978 at 18,255 thousand b/d. The second oil price rise and deceleration of economic growth depressed demand again.[42] After the 1986 oil price collapse, demand increased again and reached 17.3 million b/d in 1989.[43]

The share of oil in US primary energy consumption remained fairly stable throughout the 1970s and 1980s.[44] However, it is clear that the oil price fluctuations did influence the composition of primary energy consumption.[45]

[41] In 1970, this share was 30.9%, declining to 28.5% in 1975, to 26.7% in 1980 and to 25.9% in 1985. In 1986 and 1987, the share remained stagnant.

[42] Between 1979 and 1985, consumption decreased 16.9%.

[43] F. Niering, Oil Policy after Reagan, in: *Petroleum Economist*, February 1990, p. 59.

[44] The share of oil in US primary energy consumption was 41.8% in 1970, increased to 44.7% in 1975, declined to 43.1% in 1980, to 40% in 1985 and increased slightly again to 40.7% in 1988.

[45] See section 8.

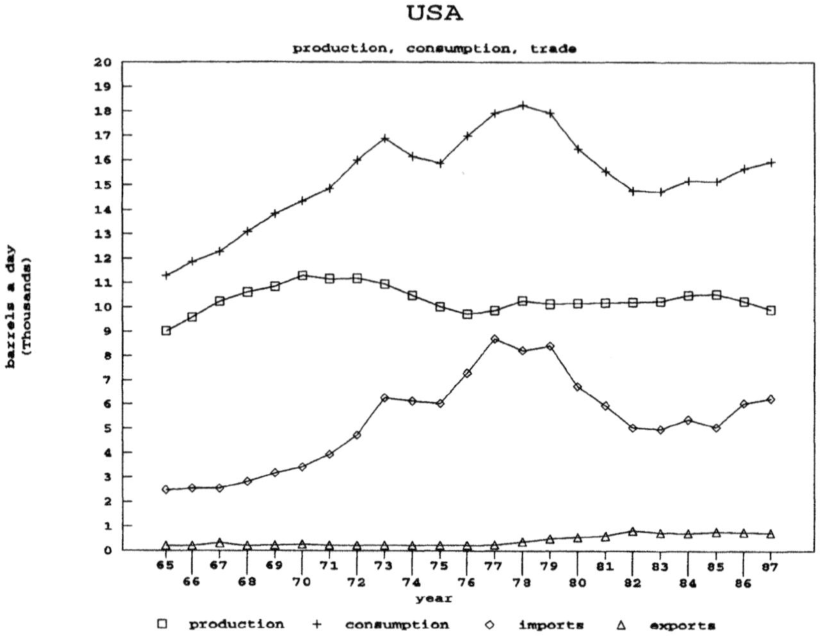

Figure 3.5

West European oil consumption increased considerably between 1960 and 1973 (See Figure 3.6). Like in the US, the oil price increase in 1973 and the economic recession temporarily depressed demand in 1974 and 1975.[46] In 1979, West European consumption reached its apex at 14,650 thousand b/d. Between 1979 and 1985, consumption declined, but it recovered somewhat after 1985. The West European rate of decline was thus slightly higher than in the US, and the rate of increase after the oil price collapse lower. The substitution away from oil, especially in the electricity generation sector, and the conservation policies seemed to have had more effect in Western Europe than in the US. The fact that the majority of the West European countries remained largely dependent on imports was an extra incentive to restrain oil demand.

West European oil consumption accounted for 13.8% of total world consumption in 1938, and this share was only slightly lower just after the war. From 1950 onward, this share began to increase steadily, reaching a quarter of world consumption in the late 1960s.[47] In the 1970s, the share in world consumption started to decline, and dropping to 20.5% in

[46] Demand declined 11.2% between 1973 to 1975.

[47] Western Europe's share in world oil consumption increased to 12.9% in 1955, to 18.8% in 1960, to 25% in 1965, to 26.9% in 1970.

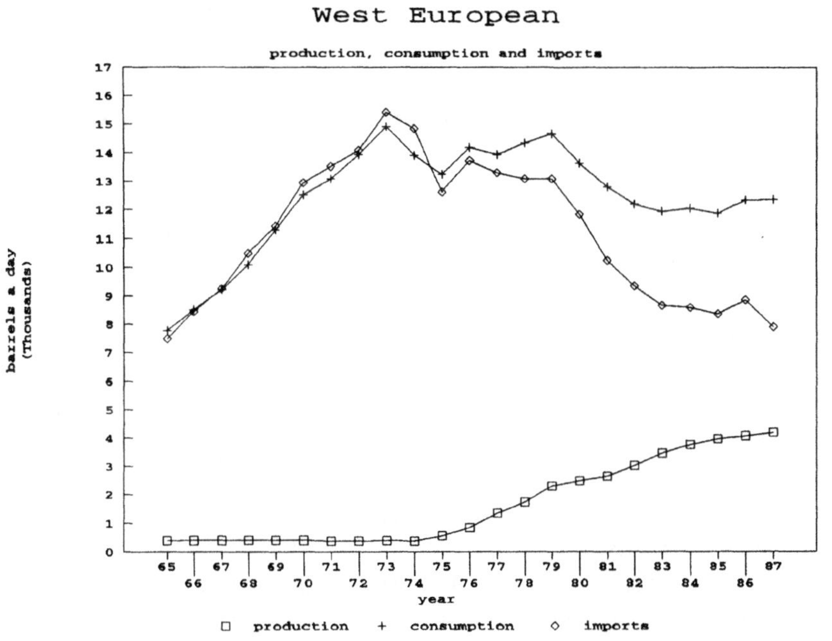

Figure 3.6

1988.

The composition of primary energy consumption in Western Europe went through profound changes after 1945. Until 1973, a major substitution process of oil for coal took place, which increased the dependency on imports. After 1973, the composition of primary energy consumption changed again, although less dramatically. The main change was some substitution away from oil.[48] For the West European countries as a group, import dependency declined notably when North Sea oil production and diversification policies took effect. However, the differences between the individual countries remained vast.[49]

Oil consumption in the US and Western Europe, as a share of world consumption and as a share of primary energy consumption, has through the years developed towards similarity. Oil demand per capita was, however, very divergent. Both oil consumption and energy consumption per capita were twice as high in the US in the early 1980s as they were in Japan

[48] The share of oil in primary energy consumption increased from 56.2% in 1970 to 56.9% in 1975, declined to 53.3% in 1980, to 45.4% in 1985 and increased to 45.6% in 1988.

[49] R. Lefeber and J.G. van der Linde, op.cit., p.452-460.

and Western Europe.[50]

Per capita energy consumption in the US was also averaging above most industrialized countries' per capita energy consumption. In 1988 per capita energy consumption in Western Europe (3.2 tonnes oil equivalent; t.o.e.) and Japan (3.3 t.o.e) was only half that of the US (8.0 t.o.e.). Compared to per capita consumption in 1978, Western Europe stabilized, the US decreased and Japan increased its per capita consumption.[51] Per capita energy consumption in the OECD countries was, however, still much higher than in LDCs and CPEs. In 1988, energy consumption per capita (in tonnes oil equivalent) in the LDC's was 0.5 t.o.e. and 1.6 t.o.e. in the CPEs.[52]

The share of imports in US oil consumption increased until 1977 (See Table 2.8) but declined from 1978 until 1985, when domestic production, in particular from Alaska, increased. In 1985, US production started to decline and the shortfall had to be made up from imports. In 1989, the total import volume was 43.9% of overall US demand and is expected to increase still.[53]

West European oil consumption was largely covered by imports until 1982. However, the North Sea oil production made the United Kingdom and Norway reasonably large net-exporters, while the majority of the West European countries remained nearly completely dependent on imports for oil consumption. In 1988/1987 West European imports increased 15.2%, bringing the share of Western Europe in world oil imports up to 33.2%, compared to a US share in world oil imports of 24.7% and a Japanese share of 16.1%. This left only a share of 26% of world oil imports for the remaining importing countries in the world.

In 1957, Japan accounted for 1.8% of world oil consumption. The rapid economic development of Japan is expressed in the growth of oil consumption (See Figure 3.4). Oil consumption in Japan expanded rapidly between 1960 to 1973. Like in Western Europe and the US, consumption contracted between 1973 and 1975 (8.1%), but increased again from 1976 to 1979 (9.3%). From the peak consumption level in 1979 (5,485 thousand b/d), consumption, as in the US and Western Europe, decreased in the period 1979-1985 (20.1%) and increased after the price collapse (8.7% in the period 1985-1988). Between 1960 and 1970 Japan's share of world oil consumption had increased from 2.7% to 8.6%, peaking in 1973 at 9.6% and decreasing slowly after that to 7.3% in 1988. Oil consumption in Japan is almost totally dependent on imports.

The share of oil in primary energy consumption in Japan was very high compared to the US and to Western Europe as a whole. However, some Western European countries

[50] In 1982, oil demand per capita in the United States was 21.48 barrels oil equivalent (b.o.e.), compared to 10.02 b.o.e in Western Europe, 12.67 b.o.e. in Japan, 3.26 b.o.e. in the CPE's and 1.55 b.o.e. in the LDC's. The average oil demand per capita in 1982 in the world was 4.4 b.o.e. and 14.22 in the OECD. S. Sinclair, op.cit., p.35.

[51] *BPSR 1989*, p. 35.

[52] *BPSR 1989*, p. 35.

[53] F. Niering, op.cit., p. 60.

resembled the Japanese dependence on (imported) oil. In a relatively short period of time, the share of oil in primary energy consumption increased to 73.8% in 1975. In 1980, due to diversification measures, this share had declined to 66.1% and to 54.8% in 1985. In 1988, in concurrence with developments in other markets, the share of oil increased slightly to 55.6%.

The development of demand by oil product group discloses some marked differences between the major OECD countries. The shift in oil product demand denotes a shift away from those oil products that can be substituted by other fuels. Especially fuel oil, used for electricity generation, suffered a large decline in demand.

The composition of demand per product in the US developed clearly further towards the lighter end of the barrel between 1978 and 1988. In 1978, the shares of gasoline, middle distillates, fuel oil and others were respectively: 38.1%, 26.1%, 17.8% and 18%. In 1988, the respective shares were: 42.2%, 28.6%, 9.2% and 20%. The dominance of gasoline and especially the drop in the share of fuel oil in the composition of product demand are striking. To meet this demand, the composition of refinery output needed to be adjusted. This could be achieved by processing lighter crudes or by upgrading the refineries. The first option was predominant in the 1970s, although the larger companies started to upgrade their refineries to meet lighter demand. An upgraded refinery can handle heavier crudes and produce a lighter output than a simple refinery. In the US, the refining sector is fairly deconcentrated. Many small refineries in the US could only handle light, sweet crudes, and their purpose was to produce gasoline. Instead of urging these small refineries to embark on upgrading of their refineries, the government protected these refineries in the 1970s by issuing subsidies and guaranteeing small and independent refiners access to crude at 'equitable' prices.[54] In the 1980s, it became clear that refinery input and output of these small companies no longer fitted the changing demand pattern. While the major oil companies had already embarked on a major restructuring process, the small refineries had difficulty accumulating the necessary capital requirements to upgrade their capacity, and their restructuring process was delayed. This put even more pressure on the small industries. The small refineries produced too much, hard to market, fuel oil, and were unable to produce unleaded gasolines. This led to some concentration in the refining sector.

The composition of product demand in Western Europe, although it also shifted to the lighter end of the barrel, is still heavier than in the US. In 1978, the shares of gasoline, middle distillates, fuel oil and others were respectively: 20.5%, 35.8%, 31% and 12.7%. In 1988, these respective shares were: 24.8%, 40.4%, 20.4% and 14.4%. The share of fuel

[54] Asked what the effect of government intervention in the refining industry had been, Harold D. Hoopman, President of Marathon Oil, answered: "Government is allocating crude to inefficient plants, plus a subsidy to keep them alive. As a result, we're probably wasting one and a quarter million barrels of oil a day." Orin Akins, chairman of Ashland Oil, a company that sold off its crude oil producing properties, answered to the same question: " (...) that an assured supply of crude is no longer essential, because in periods of shortage government will step in and allocate it." *Fortune*, January 12, 1981, p.39.

oil, in the past a major input in electricity generation, diminished substantially, but the dominance of lighter products was less pronounced. The restructuring process in the refining industry started in the late 1970s and has not been completed yet. The major companies have progressed in restructuring and rationalizing their refinery capacity at greater speed than the smaller and state-controlled refineries, leaving the structural overcapacity intact.[55] The share of the major oil companies in refining has declined since the 1950s, when they held a share of 65% of total refinery capacity, to about 40% in the 1980s. The state-controlled oil companies held substantial shares in some European countries (France and Italy) and the Independents account for approximately 30%. Just like in the US, government intervention establishes the pace and the extent of restructuring. The strategies pursued by the major oil companies are more in line with international product market developments and are opposed to the protectionist regionalistic government policies.[56]

The composition of product demand in Japan in 1978 and 1988 discloses a dominance of middle distillates and fuel oil. Although the share of gasoline increased from 17.4% in 1978 to 20.9% in 1988, the shares of middle distillates (22.9% in 1978 and 34.5% in 1988) and fuel oil (47.4% in 1978 and 27.8% in 1988) are still very large compared to the shares of these products in the US and Western Europe. The shift towards lighter products has not really materialized in Japan. This is partly due to the fact that Japan produces hardly any indigenous energy, and was not able to switch to coal and gas as quickly nor on the same scale as Western Europe.

The development of oil consumption in the group of the "rest of non-centrally planned economies", comprising mainly developing countries in Latin America, Africa, the Middle East and Asia, experienced dissimilar developments to the OECD countries. Oil consumption (See Figure 3.7) continued to grow after the two price increases, although at a somewhat slower rate. Since their level of oil consumption is generally quite low, the scope for energy conservation is limited. The share of this group of countries in world oil consumption has increased from 13.3% in 1967, 20% in 1986 and declined to 17.1% in 1988. Comparing the development of the shares in world oil consumption reveals a decrease from 71.1% in 1967 to 56.4% in 1988 for the OECD countries, an increase from 15.6% in 1967 to 21.9% in 1988 for the centrally planned economies and the mentioned increase for the developing countries.

The share of oil in primary energy consumption was, in general, in these countries substantive. In 1988, the share of oil in South Korea was 45.6%, in Taiwan 49.6%, in 'other South East Asian' countries 69.4%, in Latin America 53.6%, in the Middle East 69.5%, in Africa 42.2% and in South Asia 48.8%.[57]

[55] W. Molle, Oil refineries and petrochemical industries: moving into maturity, in: H.W. de Jong (ed.), *The structure of European Industry*, second edition, Kluwer Academic Publishers, Dordrecht, 1988, p. 52.

[56] J. Boers, G. Eisenloeffel and C. van der Linde, op.cit., p.190-195.

[57] *BPSR 1989*, p.34.

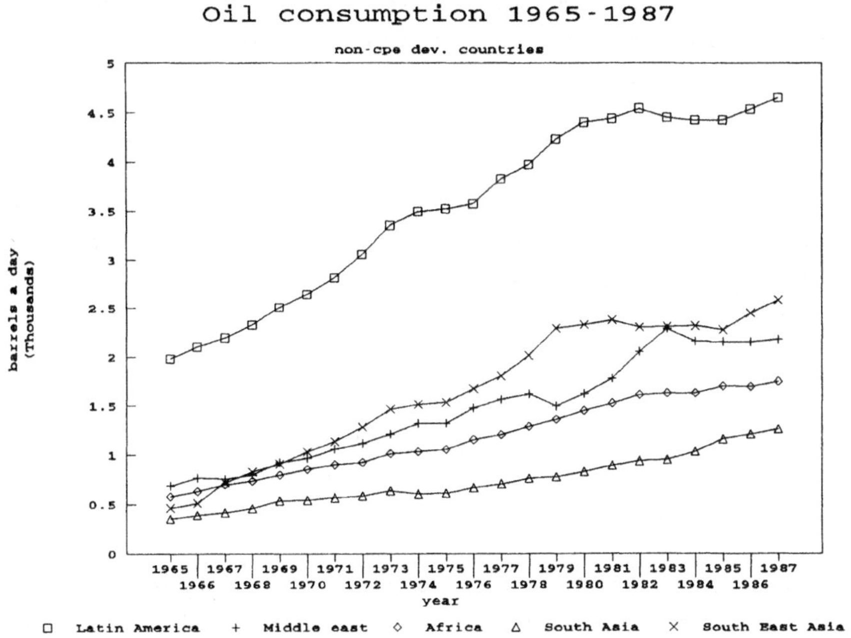

Figure 3.7

The energy intensity of production is still increasing in these countries. The pattern of international production is still developing towards labour and energy intensive production in the developing countries and knowledge and capital intensive production in the OECD countries. The pattern of foreign direct investments from the OECD countries confirms this development.

The level of industrialization and economic development is reflected in the development of the energy and oil intensity of the economy (See Table 3.2). In the OECD countries, energy intensity declined between 1950 and 1960, increased in the 1960s and declined steadily from 1970 onwards. The oil intensity in the OECD countries declined after 1973, except in 1976. In the CPEs, the energy intensity increased between 1950 and 1960, declined in the 1960s, stabilized in the beginning of the 1970s and declined waveringly from 1973 onwards. The oil intensity in the CPEs increased substantially between 1950 and 1970 and continued to rise until 1978. From 1978 onwards, the CPE oil intensity declined but was in 1982 still substantially higher than in the OECD countries. In the LDCs, the energy intensity increased almost uninterrupted from the 1950s onwards, although the rate of growth slowed down perceptively in the 1970s and 1980s. The oil intensity increased until 1971, decreased in 1972, increased again in 1973 and decreased (except in 1976 and 1979) from 1974 onwards. In 1982, the oil intensity equalled the intensity of 1970. In the OPEC countries, the energy and oil intensity were, understandably, increasing, since the industrialization efforts of these countries were geared towards energy intensive industries, like refining, petrochemicals, steel and aluminum industries.

Table 3.2 Energy and Oil intensity

	OECD		CPE		LDC		OPEC		World	
year	energy	oil	energy	oil	energy	oil	energy	oil	energy	oil
1950	6.61	2.11	12.51	1.76	2.98	1.73	1.53	1.09	6.71	2.02
1960	5.92	2.51	14.03	2.41	3.78	2.33	3.37	2.53	6.87	2.49
1970	6.11	3.25	11.20	3.08	4.52	3.08	3.19	2.17	6.73	3.19
1975	5.60	3.06	10.81	3.45	4.52	3.05	3.27	2.25	6.38	3.11
1980	5.22	2.97	10.44	3.40	4.72	3.12	3.61	2.56	6.12	2.90
1982	4.73	2.31	10.30	3.19	4.78	3.08	3.79	2.86	5.85	2.63

energy intensity= volume of energy in million barrels of oil equivalent per $1000 GDP; oil intensity= thousand b/d oil consumption per $1000 GDP; source: Stuart Sinclair, *The World Petroleum Industry*, 1984, Table 2.7, 2.8, 2.9, 2.10 and 2.11

The industrialization efforts of the developing countries and the debt problems of many of these countries make them extremely sensitive to oil price increases. Oil is a reasonably easy energy source to transport and use, given the deficient infrastructure of these countries. Development of the domestic energy sources seem limited since 91.3% of the proven natural gas reserves in 1988 are in OECD countries, OPEC countries and CPEs, and 90.7% of the proven coal reserves in 1988 are in OECD countries and CPEs. The exploration in the LDCs demands a large amount of already scarce investment capital. The development of natural gas resources in LDC's is particularly difficult. The capital requirements for exploration and production are high and the expertise highly specialized.[58] Furthermore, the introductory market for gas is restricted because of high capital requirements for transport and distribution in these countries. In general, the market would initially have to be limited to electricity generation and large-scale industries. Both the capital requirements and the uncertainty about oil price developments make the development of natural gas resources hazardous to governments. The input of capital from foreign companies has, up to now, been restricted because of the limited possibilities of exporting the gas (profitably) to other consumer markets.[59]

[58] A.R. Khan, The prospects for gas in developing countries, in: Robert Mabro (ed.), *Natural Gas, an international perspective*, proceedings of the Oxford Energy Seminar, Oxford Institute for Energy Studies, Oxford, 1986, p. 86-87.

[59] The economies of scale are very large in transport and transport enjoys decreasing per unit costs. The possibilities of transporting gas by means other than pipelines to extend the marketability of the gas are constrained by the costs incurred by liquefaction. In order to transport LNG, the gas must be cooled to -162^0 C; the LNG must then be carried by specialized transport (ship or truck) to the receiving country in a re-

The exposure to changing oil market conditions has been very large as a result. The net-exporters (OPEC and NOPEC alike) currently try to stabilize the price at $18. The oil price has become more price elastic than in the 1970s, although not for all consuming countries. The OECD net-consumers have diversified their energy consumption patterns and have invested in energy-saving. The net-consuming developing countries are price takers and less able to diversify their energy consumption pattern. Changing markets conditions will therefore have a relatively larger impact on their economies than on the economies of the industrialized countries.

6. Oil prices

From 1860 to 1973, oil prices have in terms of dollars of 1990 declined (See Figure 2.1). Before 1973, only three major price increases occurred, between 1890 and 1900, after the first world war and after the second world war.[60] The magnitude of the oil price increases in the 1970s was unprecedented.

Oil prices were since 1928 administered by the International Petroleum Cartel. Standard Oil of New Jersey (Exxon), British Petroleum and Royal Dutch Shell had agreed to a (international) market sharing and price cartel in 1928, when they concluded the Achnacarry Agreement. Other companies conceded to this agreement. From 1928 until 1944, oil prices were posted according to the Gulf Plus system.[61] From 1944, a two-point basing system was used with its equalization point first in Genoa, in 1948 in the United Kingdom and in 1949 in New York (See Chapter 4). These adjustments in the basing point system increased competition of low cost Middle Eastern oil on the American East coast market. Posted oil prices (See Table 3.3) were adjusted several times in the 1950s to balance production and demand in eastern and western hemisphere markets of the large international oil companies. The US had become a net oil importer in 1948 and demand was increasingly met by foreign oil. The price changes gave Middle Eastern oil an advantage on the American market over other crudes, in particular certain high cost American crudes.

gasification plant. J.D. Davis, *Blue Gold: The political Economy of Natural Gas*, Allen & Unwin, Boston/Sydney, 1984, p.22-29.

[60] This does not include the price volatility in oil's first decade. See also Chapter 2.

[61] The Gulf Plus system fixed oil prices, whatever the productions costs, at the level of American oil prices in the Gulf of Mexico. Transportation costs were based on the transportation costs from the Gulf of Mexico to the port of destination, whether the oil was shipped from the Middle East, Venezuela or the Gulf of Mexico. In this way the interests of the oil industry in the United States, still the most important producer as well as consumer, were freed from competition from low cost oil production areas, and investments in the American market were protected.

Table 3.3 (Posted, official government, and spot) oil prices in US$ per barrel

year	$price	year	$price	year	$price
1938	1.13	1968	1.80	1980	26.00
1949	1.84	1969	1.80	1980	28.00
1950	1.84	1970	1.80	1980	30.00
1951	1.71	1971	2.18	1980	32.00
1952	1.71	1971	2.29	1981	34.00
1953	1.71	1972	2.48	1982	34.00
1954	1.93	1973	2.59	1983	30.00
1955	1.93	1973	2.74	1983	29.00
1956	1.93	1973	3.01	1984	29.00
1957	2.08	1973	5.12	1985	28.00
1958	2.08	1973	5.18	1986	12.97
1959	1.90	1974	10.46	1987	16.92
1960	1.76	1975	11.51	1988	13.22
1961	1.80	1976	11.51	1989	15.69
1962	1.80	1977	12.09	1990	20.50
1963	1.80	1977	12.70		
1964	1.80	1978	12.70		
1965	1.80	1979	13.34		
1966	1.80	1979	14.55		
1967	1.80	1979	18.00		

note: 1938 average US crude price; 1945-1973 Arabian Light price posted at Ras Tanura; 1974-1985 official selling price of Arabian Light; 1986-1990 Arabian Light/Dubai spot price; source: 1938-1973 Adrian Hamilton, ibid, p.9; 1973- :OPEC Annual Statistical Bulletins, OPEC Bulletins and BPSR.

At the end of the 1950s, after a period of higher rates because of the Korean war and the Suez Crisis of 1956, tanker rates were decreasing. The oil companies, with BP in the lead, wanted to express the lower tanker rates in the posted prices. The leading company wanted to strengthen its position in the market after the Iranian crisis. A price decrease of $0.10 in the US was answered by an $0.18 price decrease by BP. The other companies soon followed suit. It effectively made Middle Eastern crudes more competitive with Caribbean crudes. When the American government restrained oil imports in 1959, the companies, this time with Exxon in the lead, unilaterally reduced prices again in 1960. This reduction was a response to price undercutting by the Soviet Union on the European and Indian markets.[62] The oil producing countries were furious, convening in Baghdad to join forces against the companies. They founded OPEC in September 1960. As a result, the oil companies readjusted the price cuts to a common $0.10, but did not withdraw the price cuts altogether.[63] Throughout the 1960s, the international companies maintained posted prices at $1.80. However, from the mid-1960s discounts on the posted oil prices increased, when world supply was increasingly in excess over demand (See Table 3.1). Throughout the 1960s, the posted oil price also declined in real terms (See Figure 2.1).

In 1970, the OPEC countries successfully negotiated a price increase, which was laid down in the Tripoli and Teheran Agreements of 1971. The Geneva Agreement of 1972 allowed oil prices to be adjusted to a certain extent to reflect exchange rates and inflation. However, prices were not adjusted enough to counter the full effects of the dollar devaluation of 1971 and raising inflation rates.

In October 1973, OPEC decided to administer oil prices, and prices rose sharply as a result of a tight market in 1972 and announced production cuts by OAPEC countries at the end of 1973.[64] The oil price trend set by the international oil companies was broken. Both nominal and real oil prices increased sharply (See Figure 2.1).

The aim of OPEC market regulation is in principle different from the aims of the international oil companies. The international oil companies were vertically integrated firms, and their pricing (up to approximately 1965) served different purposes than profit maximization in crude oil production alone.[65] The OPEC countries lacked (anlack) extensive downstream activities. Income maximization could only be achieved by a collective price and production policy of crude oil to counter the inequality in price and income targets of the individual member states. The fact that the member states differ in terms of realising income maximization became apparent in the second round of price increases.

In 1979, prices were increased substantially again. The terms of trade of the OPEC countries had deteriorated between 1974 (=100) and 1979 (=81)[66]. The Iranian revolution, the subsequent loss of production combined with emergency stock-building by the OECD

[62] A. Hamilton, op.cit., p.93.

[63] Ibid., p.94.

[64] The Organization of Arab Petroleum Exporting Countries announced monthly production cuts for the duration of the Israeli occupation of the Westbank, Gaza strip and Golan Heights in 1973. The Unites States, the Netherlands and Portugal were embargoed for political reasons. The international oil companies diverted oil from other countries to these countries. Although no physical shortages occurred, the diversion of oil trade and the stock-building by consumer countries did put the market under pressure.

[65] See Chapter 1.

[66] W. Bronner, R. Falk, P. Maaskola, H. Nazuri, M. Preuss, *Ölkrise und Arabische Öllander*, Marxismus aktuell, Frankfurt am Main, 1980, p.46.

countries and the break-up of the Third Party market kindled short-term demand. In 1978, world production was in deficit over demand. Increased Saudi production in 1979, to reduce the pressure on prices, could not prevent a sharp rise in prices, which was welcomed by other member states. Prices were stabilised at a rather high level.

In 1981, the official price of Arabian Light, the OPEC marker crude, was set at $34. OPEC subsequently implemented a production quota scheme to defend this price. OPEC lowered oil prices twice in 1983 and in 1984 without avail, and kept losing its market share. In December 1985, Saudi Arabia embarked on an increased market share policy and retreated from the official OPEC price policy. In about 6 months the oil prices crashed from the official price level of $28 to $10 a barrel. Real oil prices also dropped steeply. This was partly due to a drop in the dollar exchange rate in 1985 and 1986. Throughout 1986, no official OPEC price was quoted and the price was left to the market. On average the price was $15 a barrel in 1986. In 1987, OPEC tried to re-establish it pricing policy and embarked on a production restraint policy again. Oil prices slowly recovered to about $18.

Real oil price development[67] does necessitate some relativity with regard to the impact of the increase in oil prices. Real oil prices decreased slightly in the period 1974-1979. This was mainly due to the inflationary fiscal policies of the industrialized countries, the depreciation of the dollar and the transfer of increased oil prices to consumers (also export prices). In 1974, the Group of Seven suffered a $10.8 billion deficit on the current account, but managed to reverse this deficit to a $19.8 billion surplus in 1975.[68] However, cost-push inflation especially affected the energy-intensive industries, like the steel industry, which were as a result forced to rationalise.

In 1979/1980, real oil prices increased sharply again. In dollar terms, the real oil price declined again between 1980 and 1986. However, the effect of the second oil price increase was varied among the industrialized countries. The appreciation of the dollar cushioned the effect for the US and countries in the dollar-bloc. Countries with a flexible exchange rate against the dollar faced raised oil prices in national currency terms. The oil price in ECUs continued to increase with regard to the dollar oil price until the beginning of 1985.[69] The depreciation of the dollar since August 1985 (Plaza Agreement) and the abandonment of OPEC of its price and production policy in December 1985 resulted in a sharp decrease of real oil prices also for them.

[67] Real oil price: nominal price of Arabian Light deflated by the OECD export price index.

[68] G. Bird, *World Finance and Adjustment, an agenda for reform*, Macmillan, Basingstoke, 1985, p.24.

[69] *Financial Times*, May 7, 1985, p.6.

Compared to non-fuel commodity prices[70], oil prices developed erratically. From 1980 onwards, the prices of primary commodities declined sharply, while oil prices increased. From 1985 onwards, oil prices declined more rapid than primary commodity prices, creating serious adjustment problems to the oil exporting countries. In 1987, real prices for oil and non-fuel commodities recuperated slightly but are still well below the level of 1980-1984 (50% and 32% below average respectively).

7. Trade movements 1967-1988

World crude oil trade fluctuated considerably between 1967 and 1988, reflecting the response to the oil price increases in the 1970s.[71] Trade increased from 17,140 thousand b/d in 1967 to 35,350 thousand b/d in 1979, but decreased to 24,120 thousand b/d in 1985. After the price fall in 1986 trade recovered slightly. The US, Western Europe and Japan accounted for 79.2% of world imports in 1967, and only slightly less in 1988 (74%).[72]

Figures 3.8 and 3.9 clearly show the correlation between the decrease in Middle Eastern oil exports and the diminishing oil imports by Western Europe, the US and Japan. While other regions managed to stabilise their exports, Middle East exports were halved between 1978 and 1985. West Africa also suffered a decline in its exports, while the CPEs managed to increase their exports. The aggregation of trade for Western Europe concealed the development of intra-European trade in North Sea oil in the 1980s. However, the figures do disclose the reduced call of Western Europe on Middle Eastern oil in the 1980s. In 1973, Western Europe imported 10,350 thousand b/d from the Middle East. In 1980 these imports had already decreased to 7,275 thousand b/d and had dropped to 3,554 thousand b/d in 1988.[73]

Looking at the development of Middle Eastern exports over a twenty year period, growth was rather moderate from 9,140 thousands b/d in 1967 to 11,842 thousand b/d in 1988. However, a break-down in periods before, between and after the oil price increases, does disclose distinct periods of expansion, stagnation and contraction of trade. Middle Eastern

[70] Real prices are annual average nominal prices in dollars, deflated by the annual change in the manufacturing unit value index (MUV), a measure of the price of industrial country exports to developing countries. *World Bank Development Report*, 1988, p.25.

[71] World crude oil trade expanded 90.9% in the period 1967-1973, 3.4% in the period 1973-1979, contracted 25.7% in the period 1979-1986 and increased 4.2% between 1986 and 1988.

[72] The combined imports of the United States, Western Europe and Japan increased in the period 1967-1973 by 91.2%, decreased 0.2% between 1973 and 1979, 29.7% in the period 1979-1986 and increased 6.4% between 1986 and 1988.

[73] *Basic Petroleum Data Book*, Petroleum Industry Statistics, volume X, number 1, January 1990, API, Washington D.C., Section X, Table 5.

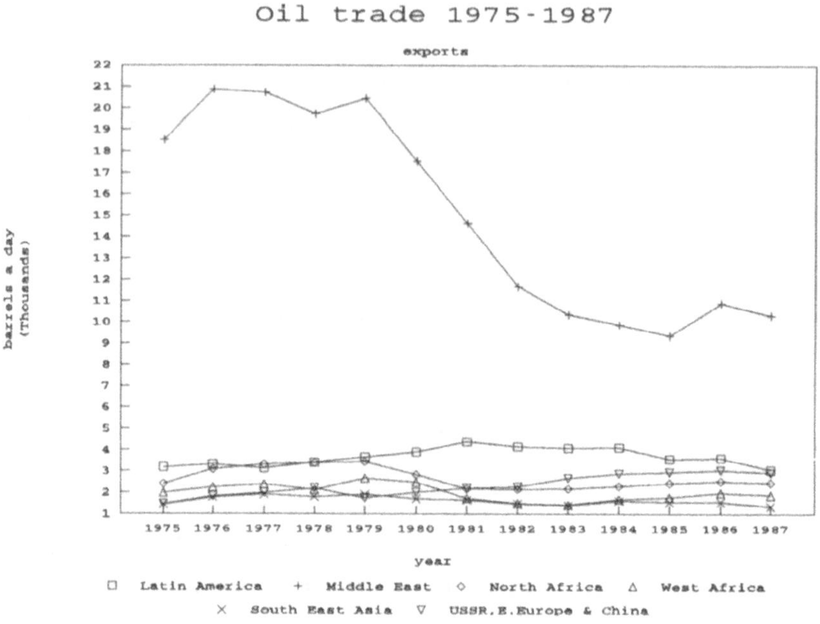

Figure 3.8

oil proved to be very price elastic. In the five years preceding the oil price rise of 1973/1974, exports expanded 118.7%. Between the first and the second oil price rise (1973-1979) exports increased only 2.2%. Following the second oil price increase, exports decreased 46.8% (1979-1986). They recaptured part of their former market share, but only after the oil price collapse. Exports expanded 16.5% in 1986/1985. In 1987/1986, exports decreased 5.2% again, when the OPEC countries made an effort to restore the oil price with production restraints. In 1988/1987, exports increased 14.8% again.

The exports of the 'rest of the world' steadily increased (except for rather minor decreases in 1974, 1975, 1980-1982) from 8,770 thousand b/d in 1967 to 15,535 thousand b/d in 1988 (77.1%), an average annual increase of 4%. In the 1980s, expansion of exports of the 'rest of the world' levelled around 14.000 thousand b/d. The oil price collapse caused exports to oscillate. In 1986/1985 exports increased 4.1%, decreased 7.2% in 1987/1986 and increased 8.8% in 1988/1987 again.

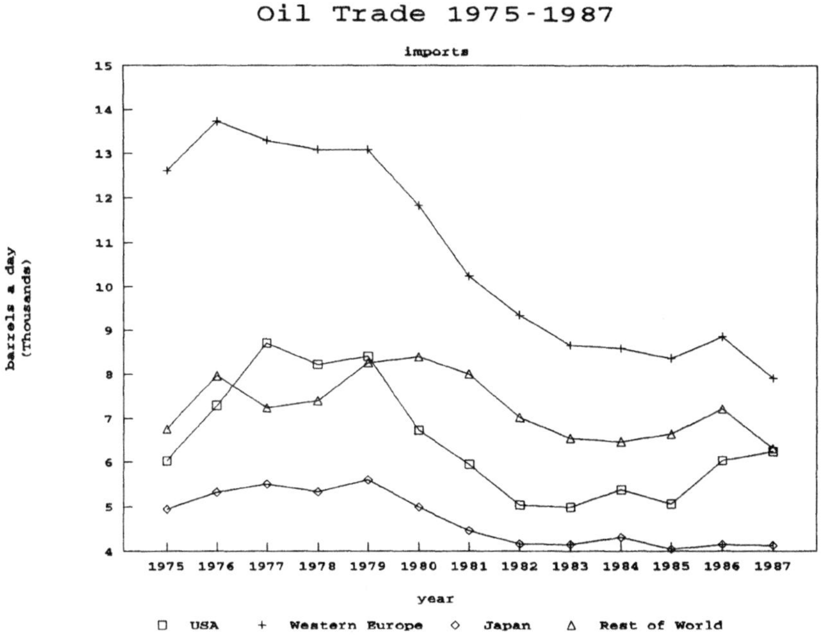

Figure 3.9

8. Diversification to source after 1973

The oil price increases in 1973/1974 and 1979/1980 undeniably had an impact on the consumption patterns of consuming countries. Governments of oil consuming countries, with a varying degree of vigour, implemented policies with the aim to substitute oil for alternative fuels, where economically and technically possible.[74]

The rate of decrease in oil consumption was by no means offset by a similar increase in coal and gas consumption (See Figure 3.10). Consumption of coal in OECD countries was fairly static in the 1970s but started to increase after 1979. Gas consumption was static in the 1970s, but actually decreased in the 1980s. The linkage of the gas and oil price, while substituting dependence on oil imports by a dependence on gas imports, appears to have been a drawback for increasing gas consumption.

Electricity generation especially experienced a reasonable rate of substitution after 1973.

[74] R. Lefeber and J.G. van der Linde, op.cit., p.452-463.

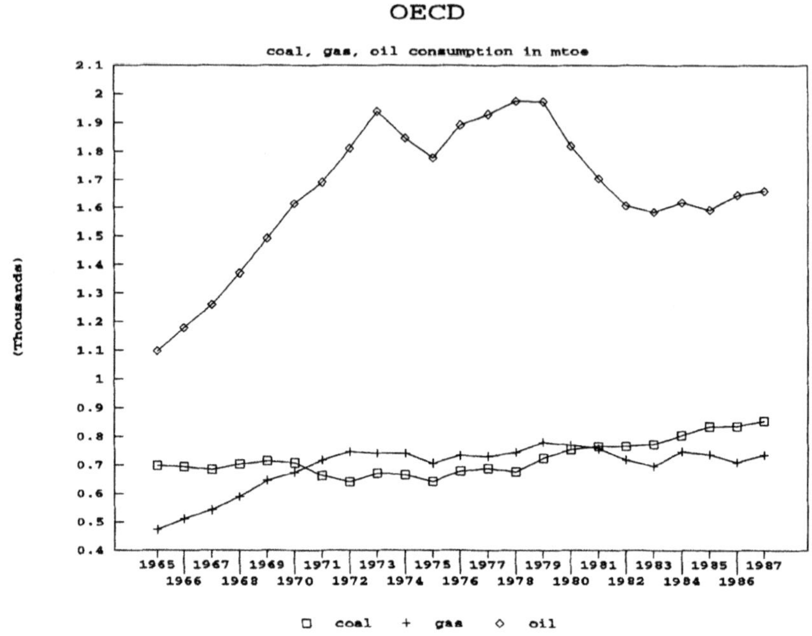

Figure 3.10

Some countries, like France and Belgium, speeded up their nuclear energy programs; other countries increased their coal and to a minor extent their gas consumption. The percentage shares of fuels used in electricity generation changed mainly in favour of coal. In 1978 the shares in the OECD countries were: oil 21.7%, gas 10.65%, coal 37.3%, nuclear energy 10.6% and hydro 19.8%. In 1988 these shares were: oil 8.6%, gas 9.2%, coal 42.4%, nuclear energy 22.45% and hydro 17.4%.[75] The decrease in the contribution of oil to electricity generation is remarkable. The increase in consumption of nuclear power and coal clearly demonstrated the extent of diversification of energy input in electricity generation stations. OECD demand in the electricity sector increased from 1151 million t.o.e. in 1978 to 1464 million t.o.e., an increase of 24.3%.[76] Gas consumption in market economies increased from 827 million tons oil equivalent in 1973 to 961 million tons in 1988, an increase of 16.2%.

Gas consumption in centrally planned economies increased from 244 million ton oil

[75] *BPSR 1989*, p.29.

[76] Ibid., p.29.

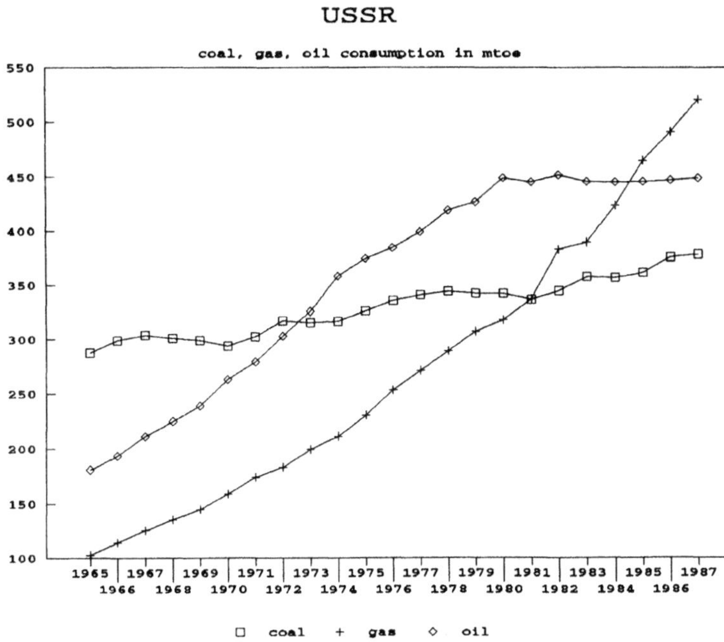

Figure 3.11

equivalent in 1973 to 664.1 million tons in 1988, an increase of 172.1%. The share of the Soviet Union in total CPE consumption is substantial, 82.7% in 1988. The fact that the Soviet Union is a major producer of coal, oil and gas, and the fact that prices deviated from world market prices, did result in a composition of consumption that developed contrary to consumption patterns in OECD countries (See Figure 3.11). Soviet gas production amounts to 40% of world natural gas production. Between 1978 and 1988, Soviet natural gas consumption increased from 289 mtoe to 549 mtoe, an increase of 90%. Oil consumption increased until 1979, and was static in the 1980s. Coal consumption increased at a much slower rate than both oil and gas.

More prominent were the changes in world coal consumption, especially consumption including CPEs. Consumption of solid fuels increased at an annual average rate of 4.3% between 1978 and 1988, while in the period 1968-1978 consumption had grown at an annual rate of 0.8% (See Figure 3.12). The energy diversification policies of the consuming

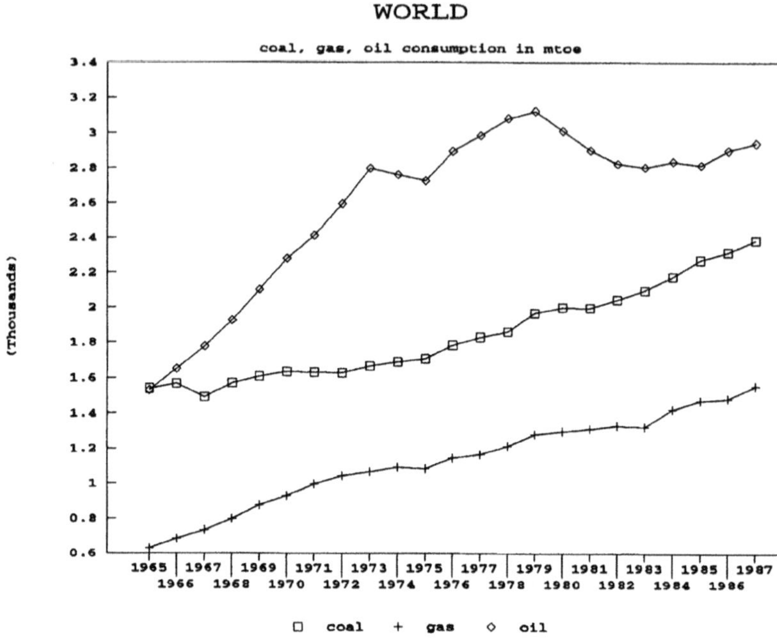

Figure 3.12

countries were, although adopted in 1974, more seriously pursued after 1980.[77] The short term price inelasticity of oil demand and the deterioration of real oil prices had left the consumption patterns throughout the 1970 largely unaffected. The second oil price rise, however, did have a stronger impact on oil consumption patterns. Demand for oil slackened and was mainly substituted by coal.

In the US, also a major producer of gas, coal and oil, consumption of coal increased moderately in the 1970s, and at a faster rate in the 1980s (See Figure 3.13). The consumption of gas declined from 1971, sharply between 1971 and 1974, was more or less stagnant between 1974 and 1979, and declined, except in 1983, in the first half of the 1980s. The share of oil in US primary energy consumption fluctuated in the period 1970-1980 between 40% and 45%. In 1988 this share was 40.7%. The share of gas diminished in this period from 36% to 30% in 1980 and to 23.7% in 1988. The share of coal was stable at 20-22% between 1970-1980, decreased to 19.1% in 1986 and increased to 24.7% in 1988.

[77] However, the time-lag between the actual investment decision and actual production, for instance, of a coal-fired power plant, is rather long and the effect of the diversification policies delayed.

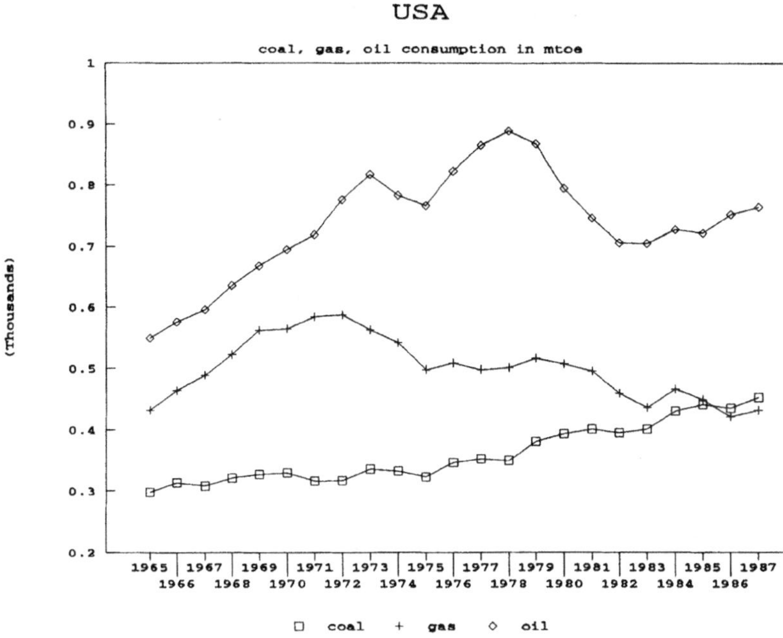

Figure 3.13

The pattern of consumption in Western Europe (See Figure 3.14) developed rather differently from the patterns in the US and the Soviet Union. Coal consumption continued to decline, but at a much slower rate, while gas consumption continued to increase, but also at a much slower rate than before. The percentage changes in the composition of primary energy consumption are mainly due to a decrease in oil consumption in the 1980s. The share of oil in primary energy decreased from 58% in 1970 to 52% 1980 and 45.6% in 1988. The share of gas increased from 7% to 13% and 15.3% in 1988. Between 1970 and 1980 the share of coal remained stagnant at 24% but had decreased to approximately 20% in the period 1986-1988.

In Japan the share of oil was close to 70% in 1970 and a little less than that in 1980. In 1973, oil consumption as a share of primary energy consumption peaked at 75%. The subsequent measures to reduce the dependency on (oil) imports resulted in a marked decline of the share of oil consumption in primary energy consumption (See Figure 3.15). In 1988, the share of oil had declined to 55.6%.

The substitution for oil by other fuels, predominantly coal, did change the composition of primary energy consumption. Coal consumption in Japan increased from 1979 onwards from

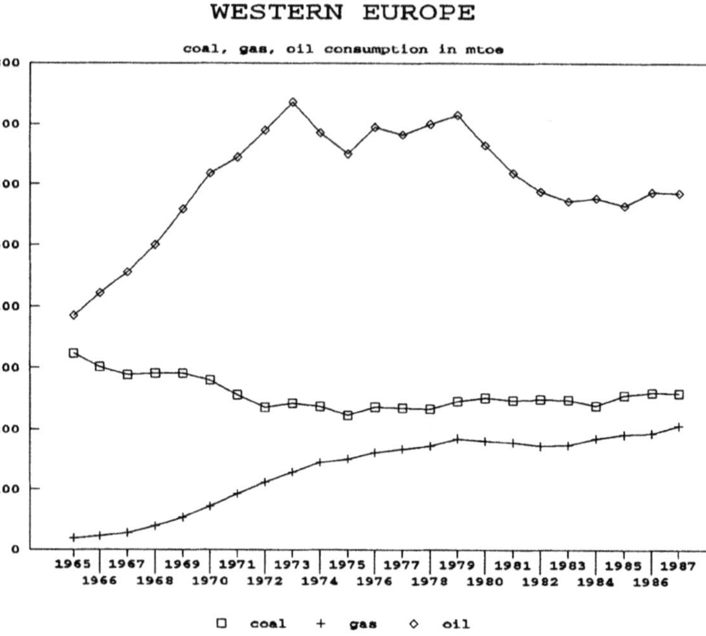

Figure 3.14

50.4 million t.o.e. to 76.2 million t.o.e. in 1988.[78] Japanese natural gas consumption increased from 15.8 mtoe in 1978 to 36.4 mtoe in 1987. Nuclear energy recorded a decline in 1988/1987 of 6.1%. The increase in primary energy consumption in 1988/1987 of 6.2% was accounted for by an increase in coal and oil consumption. The shares of coal and gas remained stable throughout the 1970s; only the share nuclear energy increased. In 1988 the shares of coal, gas and nuclear energy in primary energy consumption were respectively 19.0%, 9.8% and 10.9%.

The share of oil in the composition of primary energy consumption in the developing countries was 50.1% in 1988, while coal, gas, hydro and nuclear energy contributed respectively 21,6%, 15.1%, 11.1% and 2.1%.

China, as a major coal producer, has a dissimilar consumption pattern from other regions. The share of coal is very large, 80% in 1988, while oil accounted for only 13.9% of primary energy consumption. The Soviet Union as a major producer of oil, gas and coal, has the most diversified (hydro-carbonates) consumption pattern. In 1988, the shares of oil, gas and coal

[78] *BPSR 1989*, p.27.

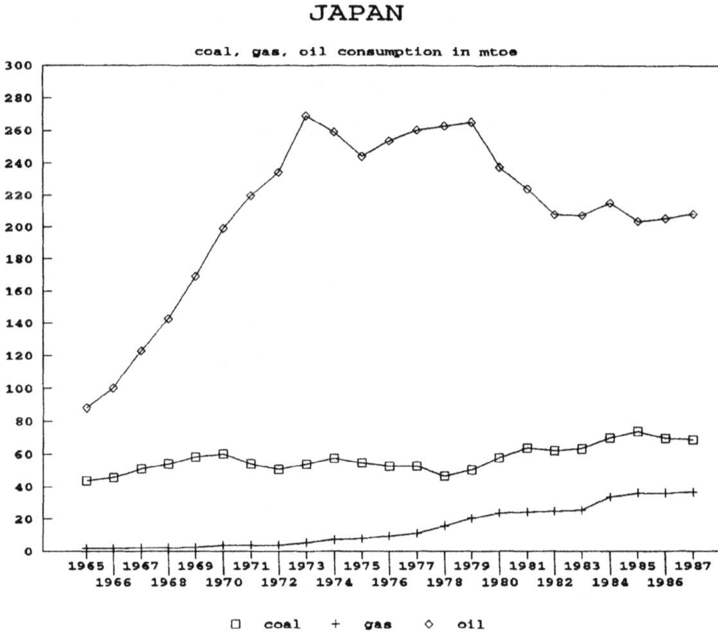

Figure 3.15

were respectively, 31.5%, 39.3% and 22.2%.

Except for China, most regions in the world are still quite dependent on oil. The composition of primary energy consumption is for most regions based on indigenous production, although production and consumption are not at balance. Obviously, Japan has a near total deficiency of production over consumption. Western Europe as a region has a smaller deficiency of supply over demand than Japan, but the distribution of production is limited to a small number of countries. Most European countries still face a near total dependence on oil imports. The US has progressively developed a deficiency in supply over demand and is major importing country now.

9. **Geographic diversification of oil production.**

The distribution of oil production in the world has changed over time (See Figure 3.16). North America was the largest oil producing region until 1969. Between 1969 and 1983, the Middle East succeeded North America as the largest producing region. From 1981 until 1986, the USSR was the largest producer. From 1986 onwards, production in the Middle

East exceeded production of the Soviet Union again. The succession by the Middle East of the Soviet Union was due to OPEC's price and production policy. The Middle Eastern OPEC members in particular have restricted their output after 1982.[79]

The share of North America in world oil production declined from 54.1% in 1950, to 38.6% in 1960, 26.7% in 1970, 19% in 1980 and 18.1% in 1988. The share of Latin America in world oil production experienced a similar development. The share decreased from 18% in 1950, to 17.2% in 1960, to 11.1% in 1970 and to 9.4% in 1980. In 1988 the share of Latin America had increased again to 11.3% of world oil production. The share of the western hemisphere production decreased from 72.1% in 1950 to 29.2% in 1988, while the share in world consumption was 70.1% in 1950 and 36% in 1988. The western hemisphere market has become more imbalanced with regard to demand and supply and thus more reliant on imports from other regions.

Western Europe, like Japan, a major consumer but endowed with little oil deposits until the discovery of the North Sea oil, has been a relatively small producer. In 1950 the share in world oil production was 6.4%, in 1960 1.4%, 0.9% in 1970, 3.9% in 1980 and 6.5% in 1988. However, the West European share is likely to diminish in the future as North Sea oil production is expected to decline. For Western Europe the imbalance between supply and demand was impressive.

Africa became a substantial producing region in the 1960s. This is reflected by the shares in world oil production which were 0,5% in 1950 and 1.3% in 1960. The discovery of oil in Libya, Algeria and Nigeria was manifested in an increased share of oil production in 1970 (13%). However, in 1980 this share had decreased to 9.8% and to 8.7% in 1988. The decrease in the 1980s is mainly due to OPEC's price and production policy, since the most important African producers are member countries.

Asia and Australasia, although comprising some substantial producing countries like Indonesia, Thailand, and Brunei, has a very limited share in world oil production indeed. In 1950 this share was 2.2%, 2.3% in 1960, 2.5% in 1970, 4.4% in 1980 and 5.3% in 1988. Supply and demand is, however, very imbalanced. The share in world oil consumption of these countries was 7.9% in 1988. The region had a deficit of supply over demand in 1988 of 1,700 thousand b/d.

The centrally planned economies have always been important oil producing countries. The relatively low share of world production in 1950 (7.7%) was mainly due to war damage. During the 1950s, oil production increased substantially and resulted in a doubled share (15.2%) in 1960. In 1970, the share increased to 16.4%, to 23.6% in 1980 and to 25.8% in 1988.

[79] See Chapter 4.

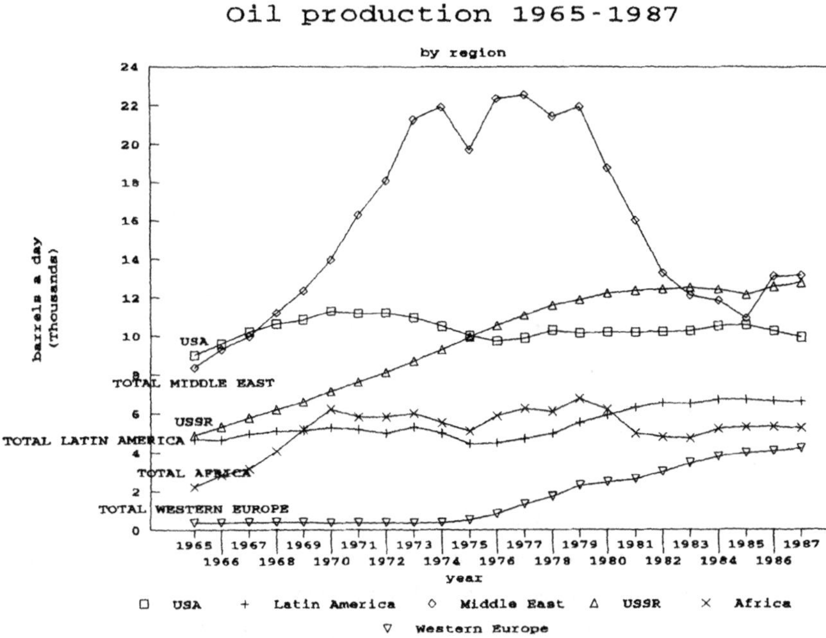

Figure 3.16

The Middle East has been an increasingly important oil producing region since the turn of the century. The development of oil production in Iran, Iraq, Saudi Arabia, Kuwait, Bahrain, Qatar and the United Arab Emirates by the major international oil companies was imperative for the control of the oil market. The low absorption capacity of most of these countries made them into the most prominent exporters. The importance of this region is expressed in the shares of total oil production. These shares were: 16.1% in 1950, 23.8% in 1960, 29% in 1970, 29.9% in 1980, 24.3% in 1988. The discrepancy between the share in world oil production and oil consumption is very striking. In 1967, the share of the Middle East in world oil consumption was 2.1%, in 1970 1.8%, in 1980 2.4% and in 1988 4.5%. The Middle East thus serves as the main supplier for crude-short regions.

The regional discrepancy between consumption of oil and production is still considerable (See Table 3.4). The share of Latin America, the Middle East, Africa, South and South East Asia in consumption was 20.2% in 1986 and had increased to 26.3% in 1988, while their combined share in production was 46.5% in 1986 and 49.6% in 1988. The share of OPEC in world oil production was 32.2% in 1986 and had increased slightly to 34% in 1988.

Table 3.4 Regional percentage shares in world production and consumption in 1988

region	production	consumption	surplus/deficit
North America	18.1	28.5	deficit
West Europe	6.5	19.5	deficit
Latin America	11.3	7.5	surplus
Middle East	24.3	4.5	surplus
Africa	8.7	2.8	surplus
Asia/Australia	5.3	15.3	deficit
(including Japan		7.8	deficit
CPE	25.8	21.9	surplus
OPEC	34.0	5.5*	surplus

*percentage share of OPEC consumption of refined products in 1987; source: OPEC Annual Statistical Bulletin 1987, BPSR 1989.

Production in the western hemisphere market (North and South America) was 4,475 thousand barrels per day short of consumption. The US imports in 1988 (6,750 thousand b/d) were comparable in volume to total Latin American oil production in that year (6,860 thousand b/d). However, the surplus of production over consumption in Latin America was in 1988 only 2,035 thousand b/d.

Production in the eastern hemisphere market (excluding CPE's) exceeded consumption 1,030 thousand b/d in 1988. Production of the centrally planned economies exceeded consumption by 2,395 thousand b/d in 1988. The Soviet Union accounted for 3,815 thousand b/d of this surplus of supply over demand. However, the surplus of consumption over production of the USSR and East Europe is becoming smaller.

The Middle Eastern role as major (future) supplier of oil for the world market is supported by the geographic distribution of world proven reserves[80] and the composition of reserves/production ratio's (r/p).[81] In 1988 the world proven oil reserves were 916.6 thousand million barrels. The regional shares in world proven oil reserves are: North America 4.8%; Latin America 13.4%; Western Europe 1.9%; Middle East 62.3%; Africa 6.1%; Asia and Australasia 2.3% and the centrally planned economies 9.2%. The share of OPEC in world proven reserves was 67.9% in 1986 and increased to 73.8% in 1988 as a

[80] Proved oil reserves are generally taken to be those quantities which geological and engineering information indicate with reasonable certainty can be recovered in the future from known reservoirs under existing economic and operating conditions. *BPSR*.

[81] Reserves/production ratio (r/p): if the reserves remaining at the end of any year are divided by the production in that year, the result is the length of time that those remaining reserves would last if production were to be continued at the then current level. *BPSR 1990*.

result of substantive new additions to reserves by Iran, Iraq and Venezuela. The OPEC had in 1986 a r/p ratio of 68.7 and, due to the new additions and still restricted output, a ratio of 94.7 in 1988. The predominant position of the Middle East in the future, with more than half of the world's proven oil reserves is evident. Within the group of Middle East countries, Kuwait (10%) and Saudi Arabia (18.6%) have a combined share of 28.6% (See also Table 5.2). The r/p ratio of both countries is also high, more than 100 years for Kuwait and 94.9 years for Saudi Arabia.[82] With regard to the development of oil consumption in the Middle East, the region will remain the most important supplier for the world market in the years to come.

The r/p ratio of North America was in 1988 10.2 (US: 9.5 and Canada: 13.6). The ratio for Latin America was 55.4, for Western Europe 12.2, for Africa 30.7, for Asia and Australasia 18.7 and the centrally planned economies 14.6. The regions with substantive r/p ratios are all currently relatively minor consumers. The share of Latin America in world oil consumption was 7.5% (4,825 thousand barrels per day), of Africa 2.8% (1,795 thousand b/d) and of South and Southeast Asia 12.3% (4,260 thousand b/d).

10. OPEC production and the importance of the Middle East

OPEC's oil production potential was in 1988 not reflected in its share of world production (34.0%) (See Figure 3.17). In 1987, the share of OPEC in exports was 45.5%.[83] This deviant picture was due to the price and production policy of the organization. After the price increase of 1979/1980, the organization tried to stabilize oil prices at $34 by restricting production. OPEC production has been reduced from the peak production level of 31,680 thousand b/d in 1977 to 21,145 thousand b/d in 1988. This low level of production compared to their production capacity and reserves (See Table 5.2) has created serious market control problems for the organization and has strained the cohesion of the organization.

The reductions in OPEC production were countered by increased production in NOPEC countries. The efforts of OPEC to stabilize the oil price at this high level were thwarted, and the oil price had to be adjusted downward several times since 1982 until in December 1985 this policy was left altogether. Saudi Arabia, for long an antagonist of a high oil price because of the fear for substitution, was not prepared any longer to be the organization's 'swing producer'. In 1985, Saudi production had decreased to 3,565 thousand b/d compared to the peak production level of 9,990 thousand b/d in 1980, when Saudi Arabia produced

[82] The proven reserves in the Neutral zone were 5.2 thousand million barrels in 1988, 0.6% of the world's proven reserves. The r/p ratio is 44.8 years. The production of this area is shared by Saudi Arabia and Kuwait.

[83] *OPEC Facts and Figures 1987*, p.21.

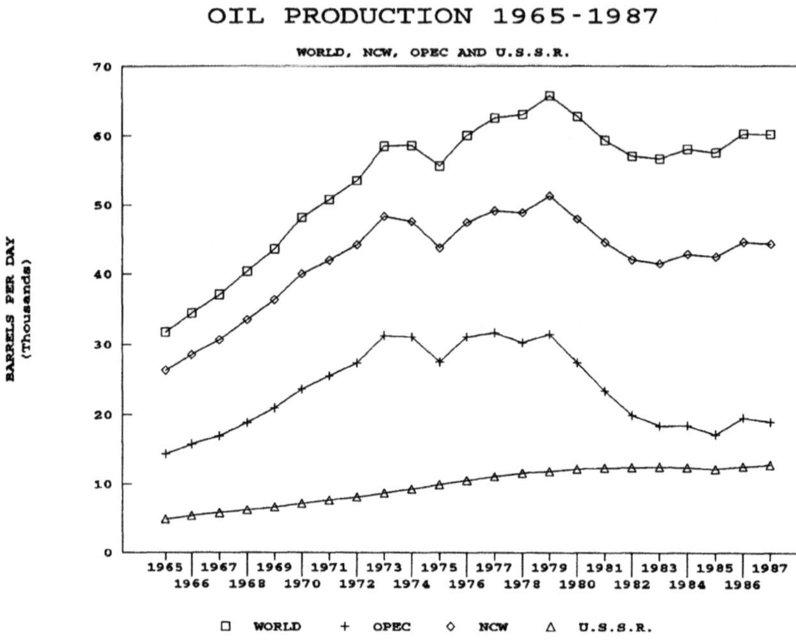

Figure 3.17

36.4% of OPEC production and 53.3% of Middle East production. Saudi Arabia's GNP at current market prices has been decreasing from $168,633 million in 1980 to $99,465 million in 1984.[84] In 1985, GNP declined to $87,626 million, a decrease of 11.9%. From 1983, Saudi Arabia also suffered a deficit on the current account and the government was forced to stall or cancel government investment programs to limit the budget deficit.

The abandonment of the official OPEC price policy in December 1985 resulted in a downward price spiral in 1986 and world oil consumption increased. Though OPEC increased production in 1986 with 2,340 thousand b/d, world production remained in deficit over world consumption. The extremely low oil price combined with a continuing depreciation of the dollar resulted in current account deficits for 10 OPEC member states.[85] However, Saudi Arabia's deficit on the current account declined as a result of increased exports. The negative economic effects of the price collapse was an incentive for OPEC to

[84] *OPEC Annual Statistical Bulletin 1987*, p.4.

[85] In 1985, Gabon, Indonesia, Iraq and Saudi Arabia had current account deficits. In 1986, Algeria, Ecuador, Iran, Libya, Qatar and Venezuela also suffered deficits. In 1986, GNP at current market prices (in million US dollars) decreased in 12 of the member states. Only Kuwait's GNP increased.

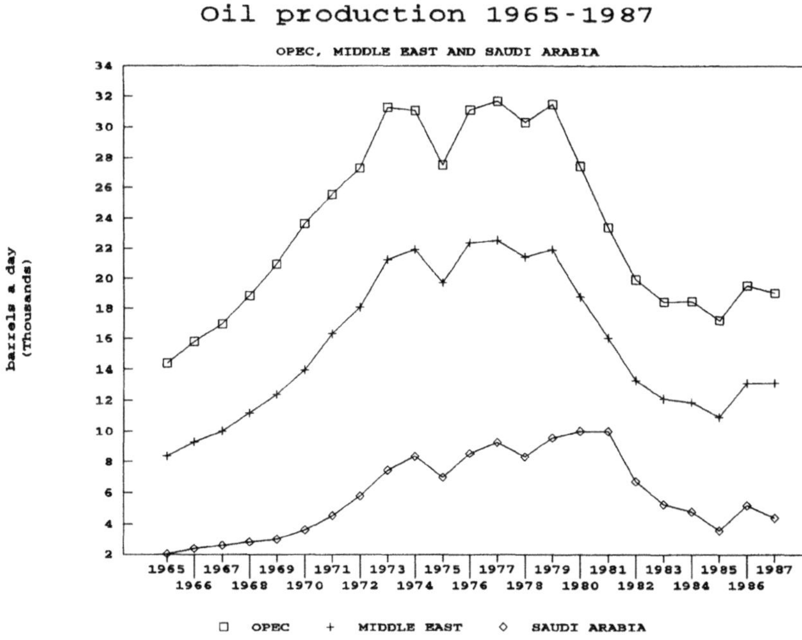

Figure 3.18

try re-establish its price policy and stabilize oil prices at $18 per barrel. OPEC, subsequently, decreased its production with 525 thousand b/d in 1987, and the deficit of world production over world consumption increased to 1,560 thousand b/d. However, spot market prices remained well below the desired $18 level. In 1988, OPEC production increased 2,115 thousand b/d and production exceeded, for the first time since 1981, 20 thousand b/d. In November 1988, OPEC agreed on a new production ceiling for 1989 of 18,500 thousand b/d.

As a result, OPEC production had to decrease 2,645 thousand b/d in 1989 compared to production in 1988. The official price of $17.52 for Saudi Arabian Light was reached in the April 1989 on the spot market, but declined on the spot market to approximately $16.5 in May and June 1989.[86]

OPEC oil production increased 171.8% between 1960 and 1988. The share of OPEC in world oil production increased from 35.3% in 1960 to 49% in 1970, decreasing in 1980 to 43.7% and to 34.0% in 1988. In the same period, Middle East oil production increased slightly more than OPEC production (187.8%), but the Middle East OPEC members

[86] *OPEC Bulletin*, July/August 1989, p.29.

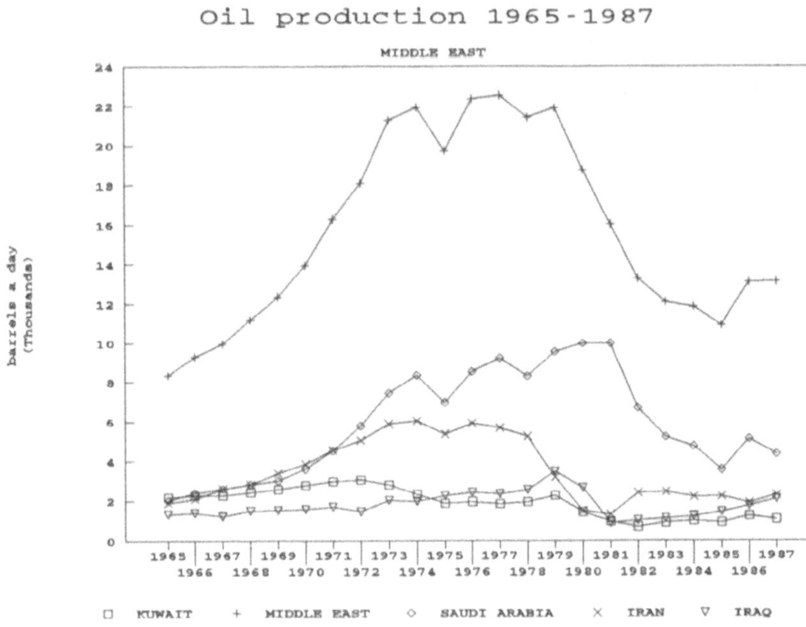

Figure 3.19

experienced lower growth of production (167.4%; See also Figure 3.18 and 3.19). Saudi Arabia's growth of production in this period was most spectacular (322.1%), despite the fact that Saudi Arabia was producing in 1988 approximately half the volume of 1980, its peak production year (See Figure 3.18).

The predominant position of the six countries around the Arab/Persian Gulf is illustrated by the share of these countries in OPEC production. In 1950 their share was 49%, and this share increased to 66.7% in 1960, decreased to 57% in 1970, increased again to 66.5% in 1980 and decreased to 65.6% in 1988. Within OPEC this group of countries, where 82.7% of OPEC reserves are located (and 61% of world oil reserves), will increasingly put its mark on policies. The Middle Eastern OPEC members are the only large surplus producers, and when other producing regions will balance supply and demand for oil, the deficit regions, notably the US, Western Europe and Japan, will be increasingly dependent on imports from the Middle East- a situation that they have tried to avert for both political and economic reasons with their energy policies in the 1970s and 1980s.[87]

11. Conclusion

In 1973, a twenty-five year period of rapid expansion of international oil production and consumption came to an end. The transition from expansion to maturity was accompanied by both organisational changes and structural changes in the oil market. Although there were earlier signs of these changes in the oil market, the events of 1973 made them explicit. The take-over by OPEC of price and production regulation resulted in changes in supply conditions. Demand was influenced by the increased cost of oil consumption, but also by other fundamental changes in the international economic environment. However, the change in oil market conditions influenced the organisation of the international economy, and the oil price increases did transform the competitive advantages of certain countries and industries.

Among these fundamental changes in the international economic environment, which affected the oil market, was the collapse of the Bretton Woods exchange rate system in March 1973, which reshuffled monetary relations in the world. Another major influence on the market was the manifestation of a new international division of labour. The post-war concentration of (international) companies on turnover as a source of growth and profit-making shifted to supply-management and the costs of production. New process-technology enabled companies to deconcentrate production and to re-locate parts of the production process to regions where specific factors of production (constituting a major cost for that part of the production process) were cheap. Aided by advanced communication technology, the deconcentration in production did not necessitate a similar deconcentration of decision-making. The increase in energy costs in the early 1970s merely added another factor to take into account and quickened the pace of these adjustments in the organisation of production. The restructuring of industry, which at first involved predominantly the transfer of labour-intensive parts of the production process to cheap labour countries, also comprised the rationalisation and, to some extent, the transfer of energy-intensive production processes. The subsequent changes in the composition of oil demand and the changes in the structure of supply characterised the 1970s and 1980s.

The oil price increases of 1973/1974 did arrest growth of oil production and consumption between 1973 and 1975, when world oil consumption declined 2.3% and world oil production 4.5%. The growth of oil consumption (14.8%) and production (18.7%) in the

[87] R. Lefeber and J.G. van der Linde, op.cit., and J.G. van der Linde and R. Lefeber, International Energy Agency Captures the development of European Community Energy Law, in: *Journal of World Trade*, vol. 22 no. 5, October 1988.

CPEs compensated in part for the drop in supply and demand in the NCW economies. In the period 1973-1975, oil consumption in the market economies declined 5.8% and oil production 9.2%. In 1974, production (both including and excluding CPEs) was in a substantial surplus over demand, but this turned into a deficit in 1975. The economic recession in the industrialized countries, aggravated by the increased oil prices, was the main cause of declining demand. When economic growth recovered in 1976, and after part of the oil price increase was inflated away, demand resumed its growth again until 1979.

However, the recovery of demand between 1975 and 1979 (15.1% including CPEs and 14% excluding CPEs) varied from country to country. Consumption in Western Europe and Japan did not resume the former upward trend like in the US and other regions in the world. In the US, oil demand was stimulated by the decreasing availability of domestic produced gas and by the government oil price controls. Moreover, since US oil production was decreasing, imports continued to increase until 1979. Demand in Latin America, South East Asia, South Asia, Africa and the Middle East stagnated in the two years after the oil price increase, but resumed their upward trend between 1975 and 1979.

This upward trend in oil consumption ceased as a result of the second oil price increase in 1979/1980. From 1979 to 1982/1983, oil demand dropped in the US and Western Europe and declined more moderately in Japan and South East Asia. Demand in Latin America continued to increase until 1982 and decreased moderately afterwards. After 1979, demand stagnated in the USSR and in Africa after 1982. In the Middle East demand continued to grow until 1983. Demand in South Asia also continued its upward trend.

From the mid-1980s onward, demand recovered again in the OECD countries, Latin America and South East Asia, while demand remained stagnant in the Middle East and the Soviet Union. Growth of consumption, however, is now much more moderate than before the oil price increases. The maturity of the main consumer markets, inter-fuel competition and the balance of payment restraints in developing countries prevented a resumption of the former upward trends.

In 1980, the previous equilibrium between supply and demand in both the world and the NCW market was disturbed. The balance in world demand and supply moved from surpluses of production over consumption to deficits. These deficits were recorded from 1981 onwards. However, refinery gains and some de-stocking and statistical errors must be taken into account. Given the level of the deficits, the world market remained fairly balanced.

The disturbance of the balance between supply and demand in the NCW market was of a more fundamental nature. The NCW market, balanced until 1980, experienced large deficits of production over consumption. Growth of CPE oil exports to the NCW market, predominantly by the Soviet Union, infringed on the market share of NCW producers. Between 1980 and 1988, the share of CPE exports in total trade increased from 6.3% to

13.9%. In the same period, OPEC's market share decreased from 76.3% to 54.8%. The increase in oil exports from the Soviet Union was not the only intrusion on OPEC's market position. Production in other NOPEC countries increased substantially in the 1970s and 1980s. The high oil price level attracted new entrants into the market. Moreover, oil produced in NOPEC countries was awarded with a premium by governments in the consuming countries.

The political impact of the oil embargo of 1973/1974 stimulated diversification as to source and as to origin of imports. The OECD governments created (tax) incentives for companies to explore and produce oil in 'politically' safe countries. The established companies as well as the new entrants needed little incentive to increase oil exploration outside the OPEC countries. The nationalization of oil fields had left the vertically integrated international oil companies crude poor, and they required new sources to restore some balance in their upstream and downstream operations. The oil price level in the 1970s and early 1980s further supported production from these marginal and sub-marginal fields.

Since the oil price collapse in 1985/1986, OPEC has gone through a difficult period. It could only compensate to a small extent for lower prices with increased production. However, production of major NOPEC sources (USA, North Sea, USSR) is either expected to stagnate or decline, and at present consumption levels the future call on oil from OPEC countries is expected to increase again. The rate of investment in exploration slackened after the oil price collapse, and substitution away from oil has been less structural than predicted. Interfuel competition has increased and lower prices stimulated oil demand. Only in specific consumer markets has oil been structurally substituted for other fuels. Oil consumption did shift towards the lighter end of the barrel in the major consumer markets. However, substitution of oil in these 'lighter' markets is less likely to occur in the short and medium term. Consequently, in order to satisfy future demand, the OPEC countries will also have to invest in the expansion of their production capacity.

Chapter Four

Development of the Middle East Oil Resources

1. Introduction

The development of the (low cost) oil resources in the Arabian/Persian Gulf area in the 1930s intensified the internationalization of the oil industry, and, from 1945 onwards, when the international market expanded rapidly, also generated a structural change in the market. A small group of Anglo/Dutch and American international oil companies dominated oil production in the Gulf area, resulting initially in an oligopolistic market structure. However, the expansionary market conditions soon attracted new comers and the market became less concentrated.

After 1945, the oil industry embarked on yet another new growth cycle. This cycle was generated by the penetration of oil in the main industrial sectors and in electricity generation first in Western Europe and later Japan. Oil also became a dominant fuel in other economies, and this internationalization of the market gave the oil industry a vital role in national economies and in the international economy. Subsequently, the industry's growing importance was matched by intensified government intervention in the market. Governments, of producer and consumer countries alike, increasingly wanted to capture the economic rents that, until then, accrued to the companies. They became important players in the market and influenced the organization of the international market.

2. The Consortia

The conclusion of the 'As Is' Agreement and the 'Red Line' Agreement[1] ('Agreements'), and the subsequent creation of the Iraq Petroleum Company (IPC) in 1928 ended a period of competition among the international oil companies. Intervention by the governments of the US, Great Britain and France in the international oil market had created favourable conditions for the companies' cartel. The isolationist foreign policy of the US after the First

[1] Within the area of the 'Red Line' Agreement the companies (party to the IPC consortium) agreed not to compete with each other for concessions nor to hold any individual concessions without seeking the permission of their partners. The 'Red Line' Agreement included the whole Arabian Peninsula, Jordan, Palestine, Lebanon, Syria and Turkey. Iran and Kuwait were not included in the 'Red Line' Agreement. C. Tugendhat and A. Hamilton, op.cit., p.84.

World War enabled France and Great Britain to increase their interests in the development of the oil reserves of the former Ottoman Empire. However, the scare over oil exhaustion in the early 1920s compelled the American government to use diplomatic force to gain entrance for American companies to the IPC. The allotment of the shares in the IPC and the 'Red Line' Agreement were the outcomes of arduous negotiations between governments and companies alike.

The formation of the International Petroleum Cartel in 1928 has dominated the development of the oil reserves of the Middle East and subsequently the international oil market. From 1926 onwards, oil prices started to decline, and this decline was expected to be aggravated by the additional supply coming on to the market. The companies envisaged a declining market cycle from the mid-1920s onwards, which could make accommodation of the market to new supply difficult, without endangering prices and profits. However, the drive of the governments of Great Britain and France to exploit the oil reserves for economic and political reasons, and the proximity to the expanding European market, made it imperative for the companies to join IPC, given the opportunity.

The oil companies, with Deterding of Shell as the main proponent, were afraid that unless the companies were able to cooperate, the combination of new supply and enduring price competition would jeopardize investments in other sectors and markets of the oil industry. Hamilton cites a letter from Deterding (Shell) to Teagle (Exxon) in 1925: "The world is now suffering from over-production, over-refining, over-transportation and- last but not least- over-retailing".[2] The conclusion of the Agreements provided the companies more or less with a guarantee of controlled exploration and production of the Middle Eastern oil reserves by the parties to the agreement.

The 'As Is' Agreement was never fully implemented with regard to market sharing. The companies suffered substantial losses during the Depression, and companies were actively competing to maintain their market shares on the American market. Between 1930 and 1935, American production increased 11%, instead of maintaining production at its 1928 level as agreed under the conservation policy, while consumption decreased 3.5%. The companies did, however, keep to their agreement in European and Asian markets.[3]

The Agreements with regard to the exploration and exploitation of the Middle East oil deposits proved more durable. However, the particular role of BP, with substantive production capacity in Iran and Kuwait, hampered the original intention of IPC to gain control. Until 1951, the production of Iran outstripped the production of Iraq, Kuwait and Saudi Arabia. The formation of the Iranian Consortium in 1953 was a determined attempt of the companies to reduce the conflicts of interests among the major companies and an

[2] A. Hamilton, op.cit., p.48.

[3] Ibid., p.49.

attempt to increase the cohesive structure of the consortia system.[4]

The limited number of (American) parties to the agreement (eventually Exxon, Mobil, Gulf, Texaco and Chevron signed the agreement), the exclusion of Kuwait, the reluctance of some of the IPC partners to expand their Middle East operations[5], and the opposition to the British dominance in the area of Saudi Arabia throughout the 1920s and 1930s[6], created opportunities for other companies to obtain concessions.[7] BP, Shell, Exxon and CFP were to witness Chevron, Texaco and Gulf begin exploration and production in the Middle East in the 1930s.

The international market conditions in the 1930s and 1940s, and the (in)ability of some of the companies to develop, process and distribute oil from their Middle East operations, were incentives for the companies to continue their cooperation after 1945, and prompted absorbtion of the new entrants in the area into the regulatory structure of the International Petroleum Cartel.

Chevron and Texaco acceded to the cartel when Exxon and Mobil became partners in Aramco in 1947.[8] The international distribution network of Chevron and Texaco, joint holders of the Saudi oil concession, was not large enough to handle the oil from Saudi Arabia.[9] The excess oil should, otherwise, have to compete on the markets of the other

[4] G. Luciani, *The Oil Companies and the Arab World*, Croom Helm/St. Martin's Press, London/New York, 1984, p.20.

[5] In the 1930s, BP was the main opponent of expanding exploration in other countries in the Middle East. The company already had access to the vast resources of Iran and later Kuwait. The other IPC partners were bound by the Red Line Agreement to obtain concessions in other countries.

[6] R. Lacey, *The Kingdom*, Fontana/Collins, 1982; *See* also D. Holden and R. Jones, *The House of Saud*, Pan Books, London, 1982.

[7] Chevron, Texaco and Gulf were not party to the Red Line Agreement and could pursue their own interests in the Middle East without prior consultation of the IPC partners.

[8] In 1933 Socal obtained the Saudi oil concession from Abdul Aziz ibn Sa'ud, king of Saudi Arabia. In 1936 Texaco became a joint concessionaire. In 1938 the California Arabian Standard Oil Company struck oil at Dhammam. In 1944 CASOC changed its name to 'American Arabian Oil Company', ARAMCO. In 1947 Mobil and Exxon became shareholders of ARAMCO (Socal 30%, Texaco 30%, Exxon 30% and Mobil 10%).

[9] "Having relied for some years on Mobil (Socony-Vacuum) to market its products in the Far East, SoCal had not itself developed any foreign marketing facilities (.....)" "On July 1, 1936, an agreement was entered into establishing a new jointly owned company, Caltex, under which SoCal received one-half interest in Texaco's marketing position east of Suez, and Texaco one-half interest in the Bahrain concession and facilities. Under this joint venture, later superseded for Saudi Arabia by the Arabian American Oil Co. (Aramco), SoCal was provided with outlets for Bahrain production by a company with a recognized market position. On its part, the Texas Company gained a more advantageous source of supply for its foreign markets. Meanwhile, SoCal also gained the concession for Saudi Arabia, to which the provisions of the Bahrain agreement were extended." "It soon became apparent that Saudi Arabia's reserves were far greater than had been anticipated and certainly far larger than could be accommodated by SoCal's newly acquired market position. By 1941, up to 200,000 barrels a day could reportedly have been produced in Saudi Arabia, whereas the available market position 'east of Suez' absorbed only 12,000 to 15,000 barrels. To compound the problem King Ibn Saud, ruler of Saudi Arabia, was insisting that production be expended to increase his royalty payments (.....)." J.M. Blair, *The Control of Oil*, Vintage Books, New York, 1978, p.36-37. "In short Aramco's production could not be readily shut in or retarded. The owners of Aramco (Standard of California and the Texas Co.) were apparently faced with the choice of either forcing their way by competitive means into markets which, before the war, had been closed

companies with the danger of flooding the market and upsetting the price structure. Gulf, another American company which operated Kuwaiti oil production in a joint-venture with BP, had similar processing and distribution problems. Between 1951 and 1965, Kuwait was the largest Middle East oil producer, and Gulf disposed of part of its surplus oil through long-term contracts with crude-short companies of the other consortia.

The interlocking relations of the companies[10] through the consortia in the Middle East (See Table 4.1) were instrumental in their dominance in the international oil industry.[11] The control mechanisms of the 'As Is' Agreement and the 'Red Line' Agreement were substituted by a control regime that gave preponderance to those with the greatest stake in long-term equilibrium.[12] The consortia arrived at the level of production by (complex) rules to reach a compromise between the highest and lowest bid. No single company could influence output to be extremely high or low, and a company that lifted more oil than it was entitled to lift had to pay a premium above cost. According to Luciani, this effectively discouraged downstream competition because "a company that was successful in increasing its share of the market downstream would be faced by an obvious tendency of all the others to restrict crude production, and would probably have to accept a competitive disadvantage (in other words, pay crude at a higher price)."[13]

The operation of the consortia enabled the companies to keep supply in line with demand, and at the same time limited competition between crude-long and crude-short companies.[14]

to them because of international cartel arrangements, i.e. the 'as is position', or permitting companies which did have marketing outlets and positions in areas west of Suez to acquire a propriety interest in Aramco." US Senate, *The International Petroleum Cartel*, 1952, p.121.

[10] The extent of the interlocking relationships between the seven companies can be illustrated by the meeting-ground of the directors of the companies in the various boards of the consortia: "(..)reveals that a considerable number of the directors of the international companies hold multiple directorships in subsidiary companies. For example, the directors of Standard of New Jersey and Socony-Vacuum, who determine the policies of Arabian American Oil Co., are the same men who help to shape the behaviour of Iraq Petroleum Co. The directors of Anglo-Iranian who assist in making high oil policy for Iraq and Iran, participate, along with the directors from Gulf, in planning the price and production policies in Kuwait. (.....) because they are directors of the parent companies, these same individuals are able to influence decisions with respect to operations in the Far East, Europe, Venezuela, and the United States." US Senate, *The International Petroleum Cartel*, 1952, p.31. Luciani remarks: "Yet the fact that representatives of the companies met at Board meetings of the IPC or the Iranian Consortium does not, by itself, prove that they were discussing their operations elsewhere and colluding in order to restrict competition." G. Luciani, op.cit., p.21-22.

[11] The output of oil in the area around the Persian/Arabian Gulf (and in other oil producing areas in the world) was controlled through joint operating companies. These companies were involved in exploration, exploitation and production of oil. Corresponding with the shares in the joint operation, sales of oil and income from their operations accrued to the parent companies. Long term supply contracts were concluded exclusively with the seven companies belonging to consortia. J.M. Blair, op.cit., p.51-52.

[12] See Chapter 1.

[13] G. Luciani, op.cit., p.23-24.

[14] Long-term contracts between the majors, under restrictive provisions with respect to competition, balanced supply and marketing possibilities for the companies. For instance, Gulf agreed to supply Shell, in a period of more than twenty years, with more than 1.5 billion barrels of oil from Kuwait.

Participation in a consortium by no means prevented the companies from exploring for cheap oil resources and exploiting these sources. The fact that the rules of the consortia penalized companies which were over-lifting, and not companies which were under-lifting, influenced investment decisions to expand production capacity outside the consortia. In fact, companies with exclusive or preferential control over oil outside the consortia could consider their consortia oil as a marginal source. Consortia oil covered their long-term needs, and other oil sources covered their needs based on their processing and distribution capacity. Since no company could predict exactly its future needs of crude oil, and since the prevailing market conditions were such that crude-short companies were penalized more severely than crude-long companies, the availability of excess production capacity was stimulated. The fact that the consortia guaranteed that this excess capacity remained idle, and would not disturb the market equilibrium, provided the individual companies with the necessary flexibility with regard to their long-term investment planning.

Table 4.1 Percentage shares of the international oil companies in the main Gulf Consortia

	IPC	KOC	ARAMCO	IRANIAN
BP	23.75	50	-	40
SHELL	23.75	-	-	14
EXXON	11.875	-	30	7
TEXACO	-	-	30	7
CHEVRON	-	-	30	7
MOBIL	11.875	-	10	7
GULF	-	50	-	7
CFP	23.75	-	-	6
Others	5.0	-	-	5

The international oil companies owned the same percentage shares in the Mosul Petroleum Company (a subsidiary of IPC), the Basrah petroleum Company, Abu Dhabi Petroleum Company, Qatar petroleum Company, and Petroleum Concessions Ltd (a company with 9 subsidiaries).

The main concern of the seven major companies was to keep a high degree of control over supply[15], stabilizing the posted oil price in such a way that the companies were able to market their 'marginal' oil from the Middle East on the world market, without displacing their other producing operations. The balance between demand and supply was a precarious arrangement, based on the fine-tuned supply regulations within the consortia. Long-term oil

[15] In 1949, the seven companies (BP, Gulf, Shell, Exxon, Mobil, Chevron and Texaco) produced 31.5% of the oil in the United States, 80.5% of oil in the rest of the western hemisphere, 99% in the Near and Middle East and 96.1% of the oil in the rest of the eastern hemisphere, excluding the USSR and Eastern European countries. US Senate, *The International Petroleum Cartel*, 1952, p.24.

contracts served to balance supply and marketing outlets of the companies, without disturbing the balance of the historical marketing positions.[16]

Mutual interest inhibited competition among the cartel members, but this could not deter other companies from competing. Oil produced by non-cartel members, like new entrants in the market, could infringe the market shares of the seven oil companies, when growth of oil production threatened to exceed oil consumption. This compelled the cartel members to curb output from the consortia to maintain the posted prices. From the late 1950s onwards, new entrants competed for markets by giving discounts on the posted price.

The post-war expansion of world oil consumption[17] did, however, facilitate a rapid development of the resources in the Middle East, although there are some noticeable differences among countries. The yearly change of production in the periods 1950-1960 and 1960-1970 were respectively: Iraq 22% and 5%, Iran 5%[18] and 13.8%, Saudi Arabia 8.5% and 11% and Kuwait 17% and 5.5%.[19] The 1960/1950 and 1970/1960 rates for world production were, respectively 7% and 8%, for the US 2.5% and 3.3%, for Venezuela 6.5% and 2.8%, for the East Indies 8.5% and 7%, for the USSR 14.5% and 9.3% and for Africa 20% and 36%.

It was noteworthy that growth of production of Iraq and Kuwait in the 1960s sank below world growth, while Saudi Arabia and Iran continued to outpace world growth. The growth of production in Africa was remarkable and emphasised the drive to expand oil production outside the consortia, on the part of both the (crude-short) consortia members and the new entrants. The constraints on output growth of the KOC and the IPC in the 1960s were possible because the relatively crude-short consortia members (Shell and CFP) were also party to the Iranian Consortium, and had already expanded their operations in Africa (respectively Nigeria and Algeria). Both BP and Gulf were traditional crude-long companies which also obtained equity oil from in the Iranian Consortium. Gulf already had to dispose of a rather substantial part of its Kuwaiti oil by the means of long-term contracts (with Shell), while the stake of BP in Iranian prevailed.[20] Exxon and Mobil had, besides shares

[16] "These two instruments of control utilized in the Middle East, joint ownership and crude-oil supply contracts, have, in effect, served to complement each other in protecting the mutual interests of the international oil companies in the production and marketing of oil." US Senate, *The International Petroleum Cartel*, 1952, p.160.

[17] Between 1950 and 1960, the average annual growth of consumption was 9.6%, between 1960 and 1970 11.5%, and between 1970 and 1973 10.8%. See Chapter three.

[18] The 1960/1955 change of production of Iran was 26%. The 1960/1950 rate was influenced by the crisis of 1951-1953. This 1960/1955 rate was for Iraq 7%, Saudi Arabia 5% and Kuwait 8.5%. *BPSR 1960*, p.19.

[19] *BPSR 1960 and 1970*, both p.19.

[20] The prevalence of BP to expand Iranian oil production and of the Aramco partners in the Saudi Arabian oil production has a political base. The nationalization of the British interests in 1951 by Mossadecq resulted in a boycott of all Iranian oil by the oil companies. In 1953, the CIA helped the Shah to take over the country, and the oil industry in Iran was subsequently operated by all companies party to IPC, KOC and Aramco. The political and strategic importance of Iran and Saudi Arabia for the United States and Great Britain resulted in a preferential treatment of these countries. To combine economic and political interests of the companies and

in the IPC, shares in the Iranian Consortium and Aramco. To Exxon, Mobil, Texaco and Socal, the Aramco concession was more important than their stake in any other consortium, because of the enormity of the Saudi reserves and the division of shares in the consortia.[21] Moreover, the good relations of the Aramco partners with the rulers of Saudi Arabia also served the political interests of the US, as the conflict over regional and religious dominance by Saudi Arabia and Iran complicated American relations with the two countries.

3. Price administration by the oil companies

After 1928, the international oil companies administered oil prices through an international single basing point system, which eliminated price differences among the sellers to the buyers of oil. The price to the buyer, including freight, was the same whether the purchase was from a nearby, low-cost source or a distant high-cost source. A base price was set and freight rates were added to this price. In order to achieve a single price for the buyers, freight costs were charged from the basing point to its final destination, irrespective of the port of shipment. The seller could take advantage of phantom freight costs, if the port of destination was closer to the port of shipment than the basing point. The aim of the basing point system was to eliminate or at least restrain price competition between the international oil companies. These companies supplied their markets from different sources of supply. Price wars to undercut another company's market position was seen as an imminent danger, and endangered investments in higher cost areas (namely the US).

The international oil companies managed to retain the basing point price system by adhering to several principles: a) the recognition of a single (or dual) basing point; b) the acceptance of the base price quotations; c) conformity to the practice of determining the delivered price on the base price; d) the adoption of common standard adjustments to differences in the quality of oil; e) the use of standard freight charges in addition to the base price.[22] These principles enabled the companies to treat oil as a homogeneous product and maintain the single price basing system, whatever the production or freight costs. The Gulf Plus system operated with a base price and basing point for freight costs determined at the ports of the US Gulf.[23]

the governments led to a division of labour in this 'oil diplomacy'.

[21] "The international supply picture is revealed as a composite of long-sustained steady increases (Saudi Arabia and Iran); of much slower rates of increase (Venezuela, Kuwait and Iraq); of precipitous rises (Libya until 1970, Nigeria, Abu Dhabi, Indonesia); and of occasional pronounced declines (Iran in 1951-54; Iraq in 1957, 1967, and 1972; Nigeria in 1968; and Venezuela and Libya since 1970). J.M. Blair, op.cit., p.101.

[22] US Senate, *The International Petroleum Cartel*, 1952, p.352-353.

[23] Ibid, p.351. The base price was based on the prices cited in 'Platt's Oilgram Price Service', published in Cleveland, Ohio. The prices cited by Platt's were questioned by the US Senate committee on their representative state. At the time the bulk of the international trade in petroleum and petroleum products were traded under long-term contracts, mostly as inter-company trade or between majors. These transactions were not part of the information on which the Platt's quotations were based.

At the end of World War II, the pricing system was modified to a two basing point system. A second basing point at the Persian/Arabian Gulf (Ras Tanura) was established. The Ras Tanura oil price was determined at the same level as the US Gulf basing point. The delivered oil price (the base oil price and freight costs) was equalized at Genoa, a mid-Mediterranean port.

After a short period, the equalization-point was shifted west. In 1948, the US had become a net-importer of oil, and the dual-basing point price system forced the producers of Middle East oil to absorb part of the freight costs, when they exported to Western Europe or the US.[24] "Thus, as long as their system remained, the companies operating in the Middle East were faced with the prospect of having to absorb freight and take a lower net price in disposing of their rapidly expanding output in the markets of western Europe and the United States. Modifications of the basing point system which would permit a freer movement of Middle East oil into the western markets seemed to be in order."[25]

The basing point system was revised, and resulted in 1948 in a new equalization point in the United Kingdom.[26] In 1949, under pressure from the European Co-operation Administration (ECA, administrator of the Marshall Plan) and several European governments, the equalization point of delivered oil prices was shifted to New York. The ECA became concerned about the fact that oil shipments to Western Europe were netting a higher price fob Ras Tanura than privately financed shipments to the US.[27] About 94% of the bulk oil shipments to Europe were financed by the ECA. The complaint of the ECA initiated a public discussion of the oil price structure.[28] In 1952 a staff report to the Federal Trade Commission of the US Senate disclosed the degree of cartelization in the international

[24] The Ras Tanura oil price fob rose to $2.22 in 1947 and enabled the companies to offset the extra absorbtion of freight (i.e. the difference between the equalising point in the mid-mediterranean and the United Kingdom).

[25] US Senate, *The International Petroleum Cartel*, 1952, p.361.

[26] "One of the early actions of the ECA (which was required under US law to ensure that suppliers charged the lowest competitive prices) was to question the validity of using the Gulf of Mexico as a western-hemisphere basing point, on the grounds that Texas had ceased to be that hemisphere's main source of internationally traded oil. The international oil companies responded by switching their western-hemisphere basing point to the Caribbean, where the prevailing fob prices for Venezuelan crudes were lower than US postings on the Gulf of Mexico by 10 cents per barrel (the amount of the US import duty on oil). Middle Eastern fob prices were then adjusted downwards during the second half of 1948 to levels which produced cif parity with Venezuelan crudes in the British market." J. Evans, *OPEC, its Member States and the World Energy Market*, Longman, Harlow, 1986, p.54.

[27] US Senate, *The International Petroleum Cartel*, 1952, p.365.

[28] "Under the companies' formula the net realized price of Middle Eastern crude imported into the United States declined when freight rates rose in 1950-51, while the fob prices posted in the Middle East remained at their end-1949 levels. The US Mutual Security Agency (ECA's successor body) subsequently accused Exxon, Mobil, Chevron and Texaco of overcharging the Marshall Aid programme by a total of $67,341,839 between mid- 1949 and mid-1952 by supplying crude oil to Western Europe at fob prices exceeding the net realized prices on the same companies' exports of Middle Eastern oil to the US market. The companies' refusal to make a further reduction in fob prices at the Persian/Arabian Gulf led to the exclusion of Middle Eastern oil from the Marshall Aid programme at the end of August 1952, thus ending the ECA/MSA's active participation in the regional export market." J. Evans, op.cit., p.55.

oil market by the Seven Sisters.[29]

The new equalization point at New York terminated discrimination in favour of Western European buyers of Middle Eastern crude, and resulted in a downward adjustment of oil prices Ras Tanura fob.[30] After the 1947 price increases (of US and Middle East oil), the price for Middle East crude decreased under pressure of the ECA to remain stable ($1.71) until 1953. In 1953 and 1957 upward price adjustments resulted in postings of $1.93 and $2.08 respectively.[31] Oil prices in the US were posted at $2.50 between 1947 and 1953 and climbed above $3.00 in 1956. Since production costs varied more than the price differentials, substantial profits were captured on Middle Eastern crudes. The average cost of production (maintenance and expansion; exclusive tax and royalty payments) in the 1950s was estimated at $0.16 in the Middle East, $0.51 in Venezuela and $1.73 in the US.[32] The price differentials between crudes from the US (and Caribbean basin) and the Middle East, enabled the companies to compete on the American market with Middle East crude. The large economic rents and profits on foreign oil production and diminishing exploration possibilities in the US attracted new oil companies in exploration and production outside the US, especially after the price equalisation point was transferred to the American East coast. The international oil companies thus secured the US market for crude oil from the Middle East at the expense of domestically produced oil.

The price structure in the 1940s and 1950s hinged on the equalization of delivered prices in the international market and was based on the relatively high cost of US oil, on freight charges that at times might not have any relationship with actual freight costs, on the control of the major companies of the bulk of tanker facilities, and on the support and maintenance by the major oil companies of world supply and demand equilibium. Thus, the basing point system, with its modifications, was essentially a complement of the production policy implemented by the companies.

4. The tax reference prices or posted oil prices and the producing countries

Throughout the 1950s, the traditional oil producing countries tried to renegotiate terms with the major oil companies. The countries wanted to increase their oil income and to re-establish some authority over their national resources. At the time, the new oil producing countries,

[29] US Senate, *The International Petroleum Cartel*, Staff Report to the Federal Trade Commission submitted to the subcommittee on Small Business, United States Senate, August 22, 1952. The research was called for on December 2, 1949.

[30] The revision of the price system to the dual basing point system prompted price increases in 1946 and 1947 for Middle Eastern oil. The prices for Arabian Light posted at Ras Tanura were: 1945 $1.05; 1946 $1.05; Dec. 1946 $1.20; March 1947 $1.60; December 1947 $2.20; July 1948 $1.99; April 1949 $1.84; July 1949-1952 $1.71; July 1953-1956 $1.93.

[31] J.M. Blair, op.cit., p.118.

[32] J. Evans, op.cit., p.62.

notably Libya, already secured better terms by applying a different concession policy[33], and by implementing (higher) taxes. In particular, the Independents were prepared to agree to these terms to secure access to foreign oil concessions.

The traditional oil producing countries were in a more difficult position. They had to renegotiate concessions and fiscal agreements with much stronger opponents with vested interests in their countries. The volume of oil that had to be processed through the integrated operations of the majors was much larger than the volumes processed by the Independents. The market was able to digest the relatively small volumes of oil outside the integrated operations of the majors at the expense of US domestic production and to supply the growing markets of Western Europe and Japan. The balance between supply and demand in the international market remained the (self-imposed) responsibility of the majors (See Table 3.1). The holdings of the companies in the Middle East played a paramount role in serving their interests as market regulators.

The traditional oil producing countries had little room to manoeuvre because they were dependent on the marketing channels of the major companies. The companies were able to switch to the (increased) output of other producing countries and successfully divide and rule among them.

The Middle East oil income was at first based on a fixed royalty per output.[34] Income was therefore positively correlated to output. The companies could reduce Middle East crude prices (as between 1947 and 1949) without impairing income. The growth of production in the Middle East after the Second World War accommodated the needs of the governments for increased development spending. Saudi Arabian production increased steeply after 1945. In 1945 Saudi production averaged 59,000 b/d, in 1946 165,000 b/d, in 1947 246,000 b/d, in 1948 390,000 b/d and in 1954 953,000 b/d.[35] Kuwaiti production experienced on average the same growth between 1946 and 1954 as Saudi production (from 16,364 b/d to 953,3301 b/d).[36] The production of Bahrain was rather stagnant between 1948 and 1954 (29,904 b/d to 30,120 b/d).[37] Iraq oil production grew less spectacular compared to Saudi and Kuwaiti production (from 97,712 b/d in 1946 to 168,586 b/d in 1951).[38]

However, the dependence of the governments of the oil producing countries in the Middle East on increased production posed problems for other oil producing countries. Especially

[33] The concession areas were smaller and allotted to a larger number of companies.

[34] Royalties were based on a barrel or ton or a constant percentage (12.5% in the Middle East and 16,66% in Venezuela) of an arbitrary price used for the purpose of accounting. This price became known as the 'posted price'. Saudi Arabia had in 1933 agreed to sell the oil concession to SOCAL for:" £35,000 down, £20,000 after eighteen months, £5,000 rental per year,£ 50,000 more if oil were discovered and a further £50,000 a year later, all in advance payments to be set against royalties of 4 shillings gold per ton." R. Lacey, op.cit., p.237.

[35] B. Shwadran, *The Middle East, Oil and the Great Powers*, Atlantic Press, London, 1955, p.343.

[36] Ibid. p.389.

[37] Ibid., p.375.

[38] Ibid., p.252.

Venezuela was worried about competition with Middle Eastern oil. From 1947 onwards, oil from the Middle East was imported into the US and competed with oil from Venezuela. The revisions of the basing-point price system had equalized prices of Middle Eastern oil on the East coast of the US and had facilitated increasing imports from the Middle East. The possibilities of the Venezuelan government to negotiate better terms in the oil industry were thus thwarted by the oil companies. In 1948 Venezuela imposed a 50% profit sharing system on the oil companies.[39] A year later, in 1949, the yearly production in Venezuela had decreased by 8 million barrels.[40] The companies increased production in the Middle East to dodge the higher acquisition costs in Venezuela.

Table 4.2 Government revenue calculation: royalty expensed versus royalty credited

in $/barrel	credited	expensed
posted price	1.80	1.80
royalty at 12.5%	0.225	0.225
cost of production	0.20	0.20
before tax profit	1.60	1.375
company profit	0.80	0.6875
government take	0.80	0.9125
of which:		
net tax (50%)	0.575	0.6875
royalty	0.225	0.225

based on: A.L. Danielsen, *The Evolution of OPEC*, 1982, p.141 and I. Skeet, *OPEC: Twenty-five years of prices and politics*, 1988, p.27.

In 1949, the Venezuelan government decided to approach the governments of the oil producing countries in the Middle East.[41] In order to diminish competition Venezuelan oil on the US market, government income in the Middle East should be made more independent of increases in production. For the Venezuelans, this would at least lessen the pressure of the governments in the Middle East on the oil companies to increase production and would reduce the possibility for the oil companies to divide and rule among the oil producing countries.

[39] Twelve days after the tax law (12 November 1948) was passed, the government was overthrown by the military. Plans to increase government control of the oil industry were shelved. P. Terzian, *OPEC: The inside story*, Zed Books Ltd., London, 1985, p.6.

[40] A.L. Danielsen, *The Evolution of OPEC*, Harcourt Brace Jovanovich, Publishers, New York, 1982, p.143.

[41] The countries of the Middle East had established the Arab League in 1945. In May 1957, the Economic Council of the Arab League adopted the idea to organise an Arab Oil Congress. The first Congress was organised in April 1959, and Venezuela was present. At the Congress the groundwork between the five founding members of the OPEC was laid.

The Venezuelan suggestion to introduce fifty-fifty profit sharing did not go wasted. Already in 1950 Saudi Arabia, Kuwait and Iraq had negotiated profit-sharing with the oil companies and by 1952 it was universally applied in the oil industry. The role of the 'posted price' was a very important one. When crudes in the 1950s were no longer priced according to US oil prices, and the companies became the sole determiners of oil prices around the world, the governments of the oil producing countries became very vulnerable. The premise of the 50-50 profit sharing agreements was a linkage of posted prices to US oil prices, a practise that prevailed at the time. The companies were the sole publishers of prices and "the only protection the signatory governments had against the possibility of a transfer of profits from the crude producing stage to another stage of the integrated operation was the established link with US crude postings".[42]

The profit-sharing agreements with the Middle Eastern countries did, however, differ with the Venezuelan system (See Table 4.2), to the disadvantage of the Middle East governments.[43] In the Middle East royalties were credited against this overall 50% tax liability, while in Venezuela the royalties were expensed. As a result, the total 'government take' of the Middle Eastern producing countries was lower than the Venezuelan government take. A comparison of Venezuelan and Saudi government revenue (See Table 4.3) illustrates the disparity in income as a result of royalty credited against taxes.

The oil companies were, under the circumstances, capable of concluding the negotiations in their own favour. On top of that they managed to obtain tax neutrality in the US. The US Internal Revenue Service ruled that revenues paid to host governments could be credited against their income tax in the US.[44] This became known as the 'golden gimmick' and had considerable impact on the oil companies' investment policy and internal accounting. The international companies shifted part of their profit margins from the domestic downstream phase of production to the (foreign) upstream stage of production. The system of posted prices was instrumental for the international companies in gaining a tax advantage over integrated oil companies in the domestic market. The posted price, on which the tax to the oil producing countries was based, did not reflect 'market prices' or 'transfer prices'. The

[42] Radical Changes in the International Oil Industry During the Past Decade, paper presented by OPEC at the Fourth Arab Petroleum Congress, Beirut, November 5-12, 1963, quoted in: I. Seymour, *OPEC Instrument of Change*, Macmillan Press, London/Basingstoke, 1980, p.27.

[43] The companies had also obtained in the agreements that taxes could not be changed. In Venezuela, taxes were not restricted and the full responsibility of the government. In 1958, Venezuela increased taxes 65-70% of profits. "The Venezuelan Income Tax Law of December 19, 1958, increased maximum income tax rates from 26 to 45% of net earnings, in addition to the flat "cedular" income tax of 2.5% which remained in effect. The other component of the government's aggregate oil income was the royalty payments, generally 16.66% of crude oil production valued according to a formula based on US Gulf prices (these being substantially higher than Venezuelan prices), continued unchanged." I. Seymour, op.cit., p.28.

[44] "The tax credit results in what public finance specialists call **tax neutrality**, meaning that a company pays the same total tax regardless of where the income is earned. (...) Governments are under pressure to adopt tax-neutral policies because companies subject to taxation in two countries cannot compete against those operating under tax-neutral laws." A.L. Danielsen, op.cit., p.142-143.

before-tax profit in the producing countries was thus enlarged and the tax liability in the home countries minimised. "For tax reasons it suited the companies very well to keep within the production stage the bulk of their overall profit from integrated operations. After the introduction of 50-50 profit sharing the majors were left with a very comfortable profit margin of 80-90 cents/barrel (something like $4.0-4.5 in 1980 dollars) at the production stage in the Middle East. This provided a very substantial pool of funds, insulated from the demands of any national taxing authority, which more than compensated for the relatively modest level of downstream returns from refining and marketing."[45] The golden gimmick proved, among other things[46], to be a stimulus for oil companies to increase or enter into foreign direct investment. The changing market conditions in the US in the 1950s were instrumental to the developments of the international market in the 1950s and 1960s.

Table 4.3 Government oil revenue of Venezuela and Saudi Arabia 1953-1962

	$ per barrel		% of gross profits	
year	Venezuela	Saudi Arabia	Venezuela	Saudi Arabia
1953	0.75	0.69	54	50
1954	0.74	0.81	53	50
1955	0.76	0.78	52	50
1956	0.82	0.78	52	50
1957	0.94	0.81	52	50
1958	1.02	0.79	65	50
1959	0.90	0.73	68	50
1960	0.83	0.70	68	50
1961	0.86	0.69	66	50
1962	0.88	0.69	67	50

source: I. Skeet, op.cit., 1988, p.29.

The US post-war policy towards (domestic) oil has been very inconsistent. The production of oil was still subject to the conservation policy of the 1930s and output restrained. Demand was increasingly met by cheap imports instead of domestic production. The golden gimmick was accepted by the IRS in 1950 for political reasons [47] but had wide economic

[45] I. Seymour, op.cit., p.14.

[46] See Chapter 2.

[47] The golden gimmick enabled ARAMCO to sooth the demands of King Saud for an increased income at the expense of the American taxpayer. "But the tax device suited the State Department as well as the King and Aramco, for it was really a means to provide foreign aid to a kingdom which was important strategically, without having to submit it to Congress: a particularly embarrassing procedure when Israel was struggling for survival. It was not until hearings six years later that the tax device was publicly aired." A. Sampson, op.cit., p.125.

consequences. Foreign direct investment in oil production soared as Independents moved abroad, attracted by the low production costs and the tax advantages. Already in 1952 the US government tried to protect the domestic industry from cheap imports.[48] However, the voluntary import restraints were ineffectual as a policy instrument. The profits on Middle Eastern oil were so attractive that imports continued to grow (See Table 2.8). In 1952 imports were 7.3% of consumption and had increased to 16.7% of consumption in 1959, when the government imposed the import-quota. The conservation policy was abandoned and oil production in the US increased.[49] Imports were controlled and increased moderately between 1960 (16.7%) and 1968 (19.9%). After 1968 the imports increased at a faster rate and accounted for 35.7% of consumption in 1973. The import-quota were especially directed against Middle Eastern crudes. However, during the 1960s the quota for these crudes was progressively enlarged as domestic production continued to fall short of production.

The immediate effect of the implementation of the import quota system of the US was a disruption of the supply policy of the international oil companies. The western and eastern hemisphere markets became separated, and demand and supply in these markets became imbalanced. The companies reduced the posted prices. The price of Arabian Light fob Ras Tanura decreased from $2.08 in 1958 to $1.90 in 1959 and $1.76 in 1960.[50] Reducing the posted prices had serious income effects for the oil producing countries. Both royalty and tax income was affected. Since the income from oil production was the main source of income of these countries, development spending by the governments was hampered. Saudi Arabia lost 9 $cents of tax income per barrel as a result of the price reduction in 1959 and 14 $cents per barrel in 1960, a reduction of income of 10% in 1959 and 8% in 1960.

5. US Government intervention in the 1950s

During the 1950s, the international oil companies progressively penetrated the energy markets in Western Europe and the US with oil from the Middle East. Consumption in Western Europe increased from 1,140 thousand b/d in 1949 to 4,060 thousand b/d in 1959.[51] The import dependency of Western Europe was extremely high (81.5% in 1949 and 92.1% in 1959). The US imported in 1949 5.5% of domestic consumption and in 1959 imports had increased to 16.7% of domestic consumption.[52]

The increased production capacity in the Middle East, combined with the international

[48] See Chapter 2.

[49] Between 1960 and 1970 oil production increased from 7,035 thousand b/d to 9,635 thousand b/d, an increase of 36%. See Table 2.8.

[50] See Table 3.6.

[51] *BPSR 1959*.

[52] See Table 2.8.

price structure and the expansion of consumption in the 1940s in the US[53], made it increasingly profitable for the major oil companies to supply the American market with foreign oil. Especially after the price equalization point was transferred to New York, oil from the Middle East could compete with domestic oil. In the US, the costs of production and the price of oil were inflated by the conservationist production policy of the government. The rising cost of exploration and exploitation of US oil impeded the competitive strength of US domestic oil companies. "For his part, Harry Truman saw growing imports as part of 'a concerted effort by the big companies to put the little [domestic] companies out of business.' He was apparently on the verge of curbing imports drastically when the Korean War broke out on June 25, 1950. The import question was left in limbo."[54]

The number of companies in the US, integrated as well as non-integrated, was extensive. The concentration in the US oil industry in the 1950s is moderately low to low[55] (See Table 4.4). Concentration in the oil production, although low on Bain's definition of concentration scales, increased after the import quota were implemented in 1958. The large international companies were forced to reassess their investment policy after the government intervention in 1958 and in order to balance their upstream and downstream activities in the western hemisphere market, increased their investments in US oil production. The increasing share was obtained either through take-overs, mergers or new exploration efforts. The inability of the companies to supply their US downstream industry with low cost foreign crude put the balance between oil production and processing capacity at odds. The declining share of the refining capacity of the top-4 and the more or less stagnant share of the top-8 points towards a policy to bring the western hemisphere market into balance again.

The combination of rising costs and increased competition from low-cost oil led to reduced exploration efforts in the 1950s in the US, and to increased efforts, especially by the 'smaller' domestic oil companies (Independents), to start exploration abroad.[56] The volume

[53] Between 1940 and 1950, U.S. consumption grew 7% while production increased 4.6%. In 1940, production exceeded consumption with 362 B.T.U., while in 1950 production experienced a deficit over consumption of 1259 B.T.U.

[54] C.D. Goodwin, Energy: 1945-1980, in: *The Wilson Quarterly*, Spring 1981, p.63.

[55] Concentration is measured by the concentration ratio. This ratio is the percentage of sales, assets, or output accounted for by a specified number of companies in the industry. The ratio in itself provides the concentration at a specific time. Taken over a period of time the ratio relates industry structure at different times. Wyant uses for the oil industry the standards of concentration as defined by J.S. Bain (Industrial Organization, 1952). Bain defined 5 types of Industry ranging from very high concentration (top 4 companies 75% or more; top 8 90% or more) to low (top 4: under 35%, top 8: under 45%). In compliance with the scale of Bain, the US oil industry between 1953 and 1972 varied from moderately low concentration to low concentration. F.R. Wyant, *The United States, OPEC, and Multinational Oil*, Lexington Books, Toronto, 1977, p.35.

[56] "From 1946 through 1955, for instance, gross investments in fixed assets and capital expenditures by oil companies within the United States were only slightly below the total gross investments and capital expenditures in the 'free world' sector of the petroleum industry outside the United States. In terms of crude oil production, crude oil runs to stills, and aggregate consumption of petroleum products, however, the "free world" sector expanded at annual rates during these years which were nearly double those experienced within the United States. What this record reveals is the fact that despite nearly equivalent expenditures on reserves, plants, and

of 'uncontrolled' oil outside the cartel increased during the 1950s as a result of the investments of the Independents in newly developed oil fields in Africa and the Middle East (especially the Neutral Zone). The large international oil companies were at that time in a difficult position to curb production (especially in Iran and Saudi Arabia), and could not restrain competition without offending the US government so shortly after the publication of the report of the Federal Trade Commission.[57] The imports also constituted a serious inroad to the rates of return of domestic producers.

Oil imports, as a percentage of total oil consumption, increased during the Truman Presidency (1945-1952) from zero to 7.3%.[58] The domestic oil production in the US grew at an average annual rate of 3.0% the 1950s, while consumption increased at an average rate of 4.9%.[59] Between 1950 and 1960, world production and consumption experienced an average annual growth of 10.2% and 9.6% respectively, almost double the growth of the American oil industry.[60]

In 1955 the US government urged the oil companies and refiners to restrain their low cost oil imports voluntarily. The policy of the US government was still aimed at conservation of US oil reserves and stabilizing the oil price by a restrictionist output policy.[61] However, as a result of this policy the profits on oil imports were too large for the companies to respond positively to this call for moderation. The smaller international oil companies could not easily reduce imports because they lacked a distribution network outside the US. Their investment efforts abroad more or less induced them to capture the profits on the imports.[62] The large international oil companies needed an outlet for the oil from the Middle East, since the companies were under pressure from the governments of Iran and Saudi Arabia to produce more oil.

equipment, there was a marked decline in the position of the American domestic industry relative to the industry outside the United States between 1945 and 1959." H.F. Williamson (et.al.), op.cit., p.817-818.

[57] The large international oil companies reacted with caution on the US Federal Trade Commission Report on the International Petroleum Cartel. "In August 1952 the United States Federal Trade Commission published a detailed and hostile report on the industry's activities during the 1930s and 1940s, and the American companies knew that any move on their part to restrict competition would lead to a disastrous clash with the Government." C. Tugendhat and A. Hamilton, op.cit., p.157.

[58] See Table 2.8.

[59] Calculations based on Table 2.6.

[60] Calculations based on Table 3.1. Oil production excluding CPEs increased at an average annual rate of 8.5% and consumption excluding CPEs at a rate of 8.4%.

[61] "Mandatory quotas did not change the underlying reality-a condition of surplus at home and abroad. A lid had long been kept on domestic production to keep oil prices stable; now a lid was clamped on imports largely for the same reason. But foreign oil was still cheaper to produce than domestic oil, and the major oil companies still stood to make greater profits by bringing it in. Controls moderated the glut but did not eliminate it." C.D. Goodwin, op.cit., p.69.

[62] "Mounting investments in foreign crude oil reserves by American companies in the post-war years increased the pressure to import crude and certain refined products for sale in the domestic market, if for no other reason than to cover a portion of overseas expenditures." Harold F. Williamson (et.al.), op.cit., p.818.

Table 4.4 Concentration ratios for the US industry

companies	top 4	top 8	top 20
oil reserves			
1970	35.5	58.5	79.4
oil production			
1955	21.2	35.9	55.7
1960	23.9	38.2	57.6
1965	27.9	44.6	63.0
1970	30.2	47.7	68.6
refining			
1947	37.0	59.0	83.0
1954	33.0	56.0	84.0
1963	34.0	56.0	82.0
1967	33.0	57.0	84.0
1972	31.0	56.0	84.0

source: Frank R. Wyant, *The United States, OPEC, and Multinational Oil*, Toronto, 1977, Table 3-1, p.36

The US government imposed a formal, but voluntary program on 1 July, 1957 and the Mandatory Oil Import Control Program (MOICP) on 11 March, 1958. Since the voluntary program had excluded restrictions on (semi)processed oil products, it had only resulted in a change of the import pattern and hardly diminished the pressure on the domestic industry. The MOICP did reduce competition, and American production increased, but this did not result in a decline in import dependence. The share of imports in consumption increased from 16.7% in 1959, to 20.2% in 1965 and 22% in 1970.[63] US production did increase more than in the 1950s (an average annual increase of 3.7% compared to 3% in the 1950s), but consumption continued to increase (4.9% in the period 1960-1970) at a higher average annual rate than production.

The change in policy was remarkable. Since 1928, conservation of US oil had been the main-stay of policy, combined with the expansion of American companies abroad. The implementation of the MOICP contradicted this former policy. The justification of this policy change was 'national security'.[64] The question whether national security was served by this

[63] See Table 2.8.

[64] J.M. Blair, op.cit., p.172-173. Blair refers to the 91st Congress, 1st Session, Senate Subcommittee on Antitrust and Monopoly, 'Hearings on Government Intervention in the Market Mechanism, The Petroleum Industry, 1969', (Part 1, p.30). Senator Hruska interviewed Professor M.A. Adelman on the governments reason for implementing the mandatory import controls. Excerpt of the interview as cited by Blair: " Sen.

change is difficult to answer. National security can be endangered when a country is physically cut-off from the necessary foreign supplies, but also by depleting the domestic oil reserves prematurely. A government seeking security of supply has the hazardous job of trying to balance a sufficient domestic production potential, while maximizing the exploitation of foreign oil supplies. The long term interests might be better served by preserving domestic oil reserves and maximizing the utilization of foreign oil supplies. However, in the short and medium term, the preservation of domestic oil reserves by limiting domestic output is usually accompanied by reduced exploration efforts and reduced exploitation of new fields, resulting in a lower level of potential production.[65]

During the Eisenhower Presidency the dependence on oil imports had risen from 7.3% to 16.7% of total consumption.[66] The swing from a restrictionist output policy to the mandatory import restrictions (and increasing output) had a considerable impact on the organization of the international oil market. The US market only absorbed incremental oil. The major oil companies could no longer compete on this market with their cheap foreign supplies, and instead had to step up their American production. The major oil companies could no longer seek to balance their upstream and downstream activities on a global scale, but were forced to balance these activities in the eastern and western hemisphere markets. They had to speed up development of the eastern hemisphere markets for their huge Middle East oil reserves. The smaller Independent oil companies continued their international expansion with even more vigour. They had to establish new marketing outlets for their foreign oil, and the expanding markets in Western Europe, Japan and the developing countries created the opportunities to circumvent the large international oil companies' control. With the assistance of governments in oil consuming and producing countries, they developed into competitors of the large companies.

6. The impact of US isolation on control of the eastern hemisphere market

The decision of the US government to restrict oil imports isolated the US market from the rest of the world. Oil companies with assets abroad needed to reassess their strategy. The situation in the western (Canada, US and Latin America) and eastern hemispheres of the international market (rest of the world, excluding centrally planned economies) differed for

Hruska: Well, the Act of 1954 as amended by the extension Act of 1955 was the basis for the President's mandatory oil import program. Dr. Adelman: Yes, sir. Sen. Hruska: And the thrust of those acts is national security. Dr. Adelman: Those are the words. Sen. Hruska: Not individual investments, not individual industries, not even the economics of the thing. The thrust was national security." Blair states that: "It is difficult to believe that the constant repetition of the words 'national security' would have been enough, in itself, to bring about such a far-reaching change in national policy, even to benefit a special interest as powerful as the oil industry." To underline that there is more, he refers to the 'Nossiter story' involving President Eisenhower's Secretary of the Treasury, Robert B. Anderson.

[65] A. Roncaglia, op.cit., p.97.

[66] See Table 2.8.

the major companies. The divergent market positions are a result of former investment and marketing policies (and for the former companies of the Standard Oil Trust- like Exxon, Mobil, Chevron and Amoco- the effects of the disintegration after the anti-trust ruling in 1911), colonial links with oil producing countries and the extent of vertical integration.

Table 4.5 Company percentage shares in NCW production and refining capacity outside the US and product sales in eastern hemisphere markets in 1953.

	production	refining	product sales
Exxon	24.9	18.9	18.0
Shell	20.6	22.3	26.2
BP	12.3	17.9	11.3
Gulf	11.2	1.2	1.5
Texaco	6.7	4.4	4.3
Chevron	6.1	3.8	4.3
Mobil	5.3	4.1	6.0

source: Neil H. Jacoby, *Multinational Oil*, New York, 1974, composed from tables 9.7, 9.9, and 9.11.

In 1953, the seven majors still accounted for 71.6% of petroleum product sales in the eastern hemisphere markets, owned 28.6% of NCW tanker capacity, operated 72.6% of NCW (excluding the US) refining capacity and produced 87.1% of NCW oil (excluding the US).[67] The shares of the majors varied substantially and stipulated already in 1953 the superiority of Exxon and Shell (See Table 4.5).[68]

In the spirit of the 'As Is' Agreement and the subsequent charters between the major oil companies in the 1930s, long-term supply contracts had been arranged to balance the market position and crude supply of the companies involved.[69] This balance was shattered when the US government implemented the mandatory import quota in 1958. Exxon, Mobil and Chevron faced deficits of crude oil production over refinery throughput in the western hemisphere and surpluses in the eastern hemisphere.

Exxon operated, after the dissolution of the Standard Oil Trust, the marketing network in Europe and Latin America but lost control over most of the crude supply in the US. In order to balance the marketing positions and the crude supply of the company, Exxon invested

[67] N.H. Jacoby, op.cit., Table 9.7, p.192; Table 9.9, p.198; Table 9.10, p.202; Table 9.11, p.206.

[68] In terms of profits (in 1949) they ranked the same as in crude oil production, except that BP, due to a lack in refining and marketing facilities, was the least profitable. In 1972 little had changed in this respect. Exxon was still producing most oil and Mobil the least of the majors, while the ranks in profits were similar to those of 1949. M. Tanzer, op.cit., p.103.

[69] Mobil obtained additional crude supplies from Kuwait under a long-term contract with BP. Shell had a long-term contract with Gulf to be supplied with Kuwaiti crude. J. Evans, op.cit., p.115 and 117.

from the 1920s onward in new oil production. These investments were increasingly abroad in Venezuela, the Middle East, Africa and in the 1970s in the North Sea. In the 1950s Exxon had a small surplus of crude oil over refinery throughput in its eastern hemisphere, but was still a crude-short company in its western hemisphere operations. Only in the 1960s, with the additional Libyan production, was the overall balance between crude production and processing more or less restored.

Texaco and Gulf were more at balance in the western hemisphere but possessed surpluses in the eastern hemisphere. Shell was in the 1950s and 1960s the largest net purchaser of crude oil of the large international oil companies. In its western hemisphere Shell had a small deficit of crude over refinery throughput, but a large deficit in its eastern hemisphere.[70] In the 1950s and 1960s, BP operations were mainly in Europe, the Middle East and the Commonwealth countries, and were less affected by the US government intervention in 1958, although they were also confronted with the difficulties of the surplus on international market.

The international oil companies did, however, face more competition from the Independents in the 1960s.[71] The new entrants in the international oil market, like Getty Oil when it obtained the concession in the Neutral Zone, had offered the governments of the oil producing countries better terms and more royalties per barrel to obtain the concessions.[72] These Independents were cut-off after 1958 from their processing and distribution networks in the US and needed to find markets for their foreign oil. According to Sampson the intrusion by the Independents in the international markets was condoned by the seven sisters: "The independents and the majors, behind all their outward battles, realised that they could not do without each other, and had reached an armed truce. The majors knew that the independents took risks in exploring which they would not take, and provided them, too, with a kind of alibi against anti-trust action. The independents used the majors to sell their oil, and needed the ordered marketing system that they had set up. It was a situation described in economists' shorthand as 'oligopoly with a fringe': or more picturesquely as being like roses growing on a pergola. The major companies, having together constructed their pergola, had an interest to maintain its stability, and they could tolerate quite well the intrusion of the

[70] J. Evans, op.cit., p.118. To cover its deficit in the eastern hemisphere, it purchased Kuwaiti crude from Gulf (with large eastern hemisphere surpluses) under a long term contract.

[71] US oil companies had expanded abroad in the 1950s to capture profits on oil imports from low-cost oil production outside the US. "The name independent is confusing, but in the international field, the word is usually applied to mean any other company other than the seven majors: that is, a company without a global system of production, transport, refining and marketing. Many of them have achieved a size and wealth which in any other industry would seem vast, so that within the United States they are regarded as majors. Some of them were originally offshoots of the Rockefeller trust, such as Standard Oil of Indiana (Amoco), Standard Oil of Ohio (Sohio), Continental (Conoco), Atlantic or Richfield, later merged to make Atlantic Richfield (Arco)." A. Sampson, op.cit., p.158.

[72] As a result, the oil producing countries in Africa implemented new concession policies that were more beneficial and allotted the concessions to more companies.

roses, which indeed made the structure look more attractive."[73]

The new entrants in the international oil industry did, however, erode the shares of operations outside the US of the seven sisters substantially (See Table 4.6). The Independents were not the only new entrants in the international oil industry. State companies in Western Europe and Japan also expanded in foreign oil production. The domestic markets of these companies (like ENI from Italy, CFP and ERAP from France and Arabian Oil Company from Japan) were protected against the operations of the other oil companies.[74]

The erosion of the companies' shares in the exploration areas was mainly due to nationalization of that part of the concessions that had, according to the host governments, not been actively explored. These areas were awarded to state oil companies of the oil producing countries and affected also the companies' share in proved oil reserves.

The erosion of market shares is for Wyant (as well as for Jacoby, who reasons among similar lines) reason to state that: "This expansion of new firms is inconsistent with the capacity or will of existing firms to exercise monopoly power. If the industry had been a controlled, powerful oligopoly it would have erected barriers to prevent the entry and growth of new companies. On the basis of this test the international petroleum industry has been quite competitive."[75] This reasoning omits an important reason for the increasingly limited power of control of the large international oil companies. The influence of public policy in consuming as well as producing countries (See Chapter 5) has been substantial in the 1950 and 1960s, when governments tried to capture some of the economic rents, and, together with increasing competition in supply, has been instrumental in changing the market conditions and eventually the market structure.

When oil became an increasingly important input in the industrialized economies in the US, Western Europe and Japan, the governments of those countries developed, at times, interests different from the international oil companies. The international oil companies had successfully replaced coal as the main fuel in these economies by offering a cheap and secure supply. The guarantee of a secure and cheap supply of oil had at the end of the 1950s already substantially changed the consumption pattern of these countries in favour of oil. The increased demand for oil products in oil consuming countries, especially in Western Europe, justified the international oil companies to switch from source refineries to market refineries. Energy demand in the Member States of the European Community was increasingly met by oil imports.[76] The coal production was initially falling short of demand, and by the mid-

[73] A. Sampson, op.cit., p.160.

[74] J. Boers, G, Eisenloeffel, C. van der Linde, op.cit., p.194.

[75] F.R. Wyant, op.cit., p.39.

[76] "During the 1950s the level of West European oil consumption increased by about 220 per cent, while the region's share of world consumption rose from 11.6 per cent in 1950 to 18.8 per cent in 1960." J. Evans, op.cit., p.23.

Table 4.6 **Percentage share of operations outside the US of the Seven Sisters and all other companies (independent and state oil companies from producing and consuming countries) between 1953 and 1972.**

	1953	1962	1972
exploration area			
seven largest	64.2	33.0	23.7
others	35.8	67.0	76.3
proved oil reserves			
seven largest	91.6	79.0	67.3
others	8.4	21.0	32.7
crude oil production			
seven largest	87.1	74.0	70.9
others	12.9	26.0	29.1
refining capacity			
seven largest	72.6	59.0	48.6
others	27.4	41.0	51.4
tanker capacity owned or managed			
seven largest	28.6	23.0	19.0
others	71.4	77.0	81.0
sales of oil products			
seven largest	71.7	59.0	54.2
others	28.3	41.0	45.8

source: Neil H. Jacoby, op.cit., Tables 9.3, p. 180/181, 9.5, p.186/187, 9.7, p.192/193, 9.9, p.198/199, 9.10, p.202/203, 9.11, p.206/207; 1962: F.R. Wyant, op.cit., Table 3-3, p.37.

1950s continental coal production was unable to compete with American and British coal and oil.[77]

Already during the 1950s, the international oil companies had penetrated the energy markets in Western Europe (and Japan) with oil from the Middle East.[78] This penetration was accompanied by establishing oil-processing plants in the main consumer markets and the build-up of a distribution network. In 1960, the refinery capacity in Western Europe

[77] R. Lefeber and J.G. van der Linde, op.cit., p.452.

[78] The origin of 80% of the crude oil imports of Western European and Japanese refineries was the Middle East. J. Evans, op.cit., p.24.

accounted for 17% of world capacity, compared to 8% in 1950.[79] Around 1955, the refinery capacity in Western Europe equalled consumption. The governments of Western Europe stimulated these investments to lessen the burden on their balance of payments and scarce foreign exchange reserves.[80] The import of crude oil was cheaper than the import of oil products.[81] After some oil processing capacity was installed, the governments of consuming countries imposed tariff barriers or import quota on imports of oil products. The low cost Middle Eastern oil enabled the companies to compete with other energy sources and to differentiate prices of oil products accordingly. In product markets without any substitutes, the price of these products could compensate for losses in other product markets. In the 1950s and 1960s, fuel oil increasingly substituted for coal in the electricity generating sector in the Member States.[82]

The oil companies were able to put the growing demand for oil products in Europe to their benefit. Oil demand increased as economic growth picked up in Europe and oil replaced coal as the main fuel. From 1950 to 1960 oil consumption in Western Europe increased from 1,300 thousand b/d to 4,060 thousand b/d, an average annual increase of 21.2%.[83] In 1957, oil replaced coal as the main fuel in Italy. The Netherlands followed in 1962, France in 1967, and by 1972 oil was the main fuel in the United Kingdom, West Germany, Belgium and Luxembourg.[84]

The distribution of a wider range of oil products increased the productive use of a barrel of oil. As more different oil products were marketed in Europe, the transport of crude oil became advantageous, as the crude contained less waste products than before and transport of separate oil products from source refineries became relatively more expensive.[85]

[79] W. Molle and E. Wever, op.cit., p.31, Table 3.3.

[80] "Thus, the European governments used both the carrot and the stick to persuade the companies which supplied the oil to implement this policy. The companies were, for example, given loans at less than the then current rates of interest for financing the construction of refineries and they were also guaranteed the availability of foreign exchange to import the necessary equipment. At the same time, however, they were told that once the refineries had been established, then preference would be given, through tariff or quota arrangements, to crude oil imports over product imports. This, of course, meant that companies which failed to build refineries could well find themselves forced or priced out of the markets." P. Odell, *Oil and World Power*, Penguin Books, Harmondsworth, 1986, eighth edition, p.115.

[81] In 1960, already 90% of the imports of oil in Western Europe were crude oil imports. The imports of oil products had been reduced to a minimal level. J. Evans, op.cit., p.23.

[82] "The largest single product has been 'fuel oil', accounting for a half to two-thirds of total consumption. Diesel oil, closest to fuel oil in nature, takes up some 10 per cent of total consumption. Fuel oil and diesel oil are used for many purposes. They are important as fuel for electric-power generation; they also represent almost the entire consumption of the sub-sectors railways, inland waterways, and sea bunkers of the transport sector. Moreover, large parts of the industrial oil consumption are in the shape of fuel and diesel oil, and in agricultural and residential use these products account for large parts of the total consumption of oil products (oil-fired central heating, small power generators, and so on). W. Molle and E. Wever, op.cit., p.35-36.

[83] See Table 3.2.

[84] P. Odell, op.cit., p.120/121. In 1972, natural gas replaced oil as the main fuel in the Netherlands.

[85] W. Molle en E. Wever, op.cit., p.30.

Furthermore, the political situation in the oil producing countries became more hostile to the international industry. Political independence stimulated nationalist feelings in developing countries and kindled their strivings for economic independence. This prompted the oil companies to divert their investments in downstream operation in the consumer countries.

The preference of the majors to develop the market refinery capacity instead of increasing the source refinery capacity reduced the oil producing countries to crude oil suppliers instead of oil product producers. The value added accrued to the companies and the oil consuming countries, especially after the downstream industry produced an increasing number of finished products based on oil.

The US government had chosen to implement import controls to protect the domestic oil industry from cheap oil imports from the Middle East, while other governments chose to protect their domestic coal industry. Market regulations with regard to production, imports, processing and other energy sources, did alter the market conditions. The declining concentration ratios can, therefore, not only be explained by diminishing oligopolistic power of the international oil companies, but they also reflect the impact of government intervention.

The market conditions were not exclusively influenced by government intervention in the industrialized economies. The oil producing countries (old, new and potential) acquired more government control over their natural resources.[86] Concession areas in new oil producing countries were smaller, and the governments sought to diversify the number of companies exploring for oil. The most notable example of this new policy was the government of Libya in the 1960s. Libya's oil concessions were granted under the Petroleum Law of 21 April, 1955. The law was aimed at attacking a high level of investments in a short period of time (by granting the companies certain fiscal advantages during the exploration period) and attacking a large number of companies by restricting the number of concession areas that could be obtained by one operator.[87] This affected the share of the large international oil companies in the proven oil reserves, concessions held and oil production (See Table 4.6) outside the US (excluding the centrally planned economies). The traditional oil producing countries did try to curb the influence of the large international oil companies by relinquishing parts of concessions which were not explored.[88] Part of the effort of the

[86] At the Bandung Conference in 1955, the developing countries initiated a movement of developing countries that aimed to obtain political and economic independency. Later the group of countries became known in international fora as the 'Group of 77'. The Group of 77 (currently with more than 120 participating countries) prepared joint statements on issues dealt with in international conferences like UNCTAD. The sovereignty over natural resources was an important issue in the 1960s and 1970s.

[87] In the first round of licensing in 1956, 51 concessions were granted to 17 companies. Oil was found in 1957 and production exceeded 1 million b/d around 1965. The Petroleum Law of 1955 was revised in 1957 and 1961, omitting or modifying some of the fiscal advantages in the initial law. J. Evans, op.cit., p.258-259.

[88] After the overthrow of the monarchy in Iraq in 1958, the new government started negotiating with the IPC partners. Among the issues was the demand to relinquish concessions areas. As the negotiation dragged on, the government demanded, in the beginning of the 1960s, the relinquishment of 90% of the existing concession

governments to increase control over domestic oil production was to establish national oil companies.[89] However, the share of the Seven Sisters in the traditional oil producing regions remained high, although the policy of the governments of the oil producing countries did result in a growing number of companies involved in oil production in their countries. In the newly producing regions, in particular Africa, the share of the large companies is much lower (See Table 4.7).

The large international oil companies had entered a period of diminishing direct control in the 1960s. The isolation of the US market from the international market had obstructed the development of the consortia. The processing capacity in the Western European and Japanese markets, although they developed rapidly, set the pace for the output of the consortia. Government intervention had increased in consuming as well as producing countries as the oil industry, stimulated by the large companies, gained importance in the international economy. The large oil companies reacted differently to these challenges to their control.

The expansion of the large international oil companies was in the 1950s and early 1960s mainly downstream, in the (petro)chemical industry. This expansion is understandable since the companies owned the basic feedstocks for this industry, had the necessary investment capital at their disposal and were supplied by the refining industry with familiarity with the processing patterns of the chemical industry. In 1962, over one-third of the US petrochemical industry was owned or operated by the oil companies.[90] In 1947, Exxon, Mobil and Texaco ranked among the top 10 US manufacturing firms in terms of total assets. In 1973, Exxon ranked 2nd, Shell 4th, Texaco 7th, Mobil 8th, Gulf 12th, Socal 14th and BP 15th on the Fortune list of largest companies of the world.[91]

The next expansion of the major oil companies in the 1960s was into competitive fuel industries like coal and nuclear power. The growth of environmental problems in the US, Western Europe and Japan added to the oil companies' uncertainty about which energy sources would become dominant in the future. "Of the top twenty owners of coal reserves, eleven are oil companies, and these control 40 percent of total US coal reserves. As a result of this increased concentration, the eight largest uranium mining companies control 70 percent of US reserves and over 80 percent of production. Three of the four leading U.S. uranium producers are now oil companies-- Kerr-McGee, Atlantic Richfield and Exxon-- and

areas and the introduction of new concession terms with respect to unexplored areas. In June 1961 the companies offered to relinquish 75% immediately of their concession areas and an additional 15% within 7 years. In December 1961, the Iraqi government implemented Law 80, restricting the concession areas of the Basrah Petroleum Company, the Mosul Petroleum Company and Iraq Petroleum Company (see Table 4.1) to a combined total of less than 2000 sq.km. J. Evans, op.cit., p.236-237.

[89] In November 1962, the state corporation Petromin (The General Petroleum and Mineral Organization) was established in Saudi Arabia.

[90] M. Tanzer, op.cit., p.106.

[91] W. Bronner (et.al.), op.cit., p.24.

the oil companies have an even firmer grasp over uranium than coal: they own 55 percent of the reserves and undertake 60 percent of the exploration."[92]

Table 4.7 Geographic shares of seven largest and all other companies in NCW market outside US in 1953 and 1972

	1953		1972	
	seven largest	other	seven largest	other
proved crude oil reserves				
Latin America	77	23	60	40
Middle East	94	6	83	17
Far East	94	6	74	26
Africa	89	11	25	75
oil production				
Latin America	81	19	73	27
Middle East	92	8	83	17
Far East	94	6	71	29
Africa	92	8	47	53
refining capacity				
Latin America	67	33	51	49
Western Europe	65	35	55	45
Middle East	98	2	63	37
Far East	77	23	34	66
Africa	32	68	30	70

source: Neil H. Jacoby, op.cit., composed from tables 9.4, p.183, 9.6, p.188, 9.8, p.194.

The horizontal integration by the oil companies was caused by a number of changes in the international oil market.[93] The new entrants on the international market managed to capture an increasing share of the market by raising discounts on the posted price. The oil producing countries had protested against the downward adjustment in the posted price in 1960 and found in this price decrease a motive to combine forces against the large oil companies' market control and established OPEC. Although OPEC was not taken very seriously by the companies in the early 1960s, the growing nationalist feelings in the oil producing countries

[92] M. Tanzer, op.cit., p.113 and 119-120.

[93] Ibid., p.107 (except the remark on the isolation of the US market and its relation to the oil price).

made these countries less dependable with regard to long-term security of supply. The oil companies' doubt about their long-term supply position in the Middle East was an incentive to give preponderance to supply instead of steady (real) oil prices. In the 1960s, discounts on the posted price were increasing and the real price of oil declining. From 1967 to 1971, the surplus of production over consumption increased, and the control of the companies was virtually terminated.[94]

7. Conclusion

The expansionary phase of the post-war growth cycle was accompanied, like in earlier analogous periods, by a process of deconcentration of the market. New entrants infringed on the market shares of the large international oil companies that had dominated the international market before. The position of the large international oil companies remained, due to the large reserves and production capacity of the Middle Eastern resources, an important one. These companies were the operators of the Middle East consortia, functioning as swing producers in the NCW market in the 1960s. The position of the large oil companies was not only infringed upon by new entrants but also by deepening government intervention in the market.

The major oil companies had developed into vertically integrated companies on a world scale. They were integrated in all stages of production: exploration, exploitation, transport, refining and distribution. In the 1950s they moved further upmarket in the petrochemical industry. However, the level of international vertical integration, or globalisation, of their operations varied among the companies. Exxon, Shell and BP had internationalised their operations to a greater extent than the other American companies.

The replacement of the Gulf Plus system by the two basing-point system was a response to a (government-induced) changing supply and demand pattern and to the accompanying modified structure of the (integrated) divisions of the large international oil companies.[95] The companies had, with the exception of Texaco and Gulf, deficits in crude oil production over refinery throughput in their western hemisphere markets. This advantage of producing and marketing cheap Middle East crude in the US was phased out by the Gulf Plus price

[94] See Table 3.1.

[95] "The eight international majors controlled over 90 per cent of production, refining and marketing outside the United States and the communist bloc in the early 1950s. Six of them (i.e. excluding BP and CFP) together controlled nearly one-third of US crude production and about 45 per cent of total US refining capacity (but up to 60 per cent of capacity in the main east coast importing centres). The initial evolution of crude oil prices in the Middle East after the Second World War was essentially based on adjustments to the major companies' internal transfer-pricing structures, details of which were publicly known only insofar as invoiced crude prices affected cross-border trade and the division of total taxable income between different countries. The general trend of these adjustments was subsequently rationalized by the companies (in terms of the need to maintain "competitive" cif parities) when they were called upon to justify their crude oil pricing policies to the ECA, whose attention was focused on the way in which transfer-pricing (or, very rarely, contract sales to non-affiliated customers) affected import costs in Western European countries." J. Evans, op.cit., p.55.

system. The magnitude of the production capacity of the Middle East could only be used to the companies' advantage by changing the price structure.

The difference in prices of crude from the Middle East and the US gradually widened. Since production costs varied more than the price differentials, substantial profits were captured on Middle Eastern crudes. The price differentials between crudes from the US (and Caribbean basin) and the Middle East, enabled the companies to compete on the American market with Middle East crude. The large economic rents and profits on foreign oil production, and the diminishing exploration possibilities in the US attracted new oil companies into exploration and production outside the US, especially after the price equalisation point was transferred to the American East coast.

The two basing-point pricing system provided the companies with a price structure which appeared to do more justice to production costs and freight costs and to competition in the oil industry. However, throughout the 1950s, Middle East oil prices paralleled the price changes of US Gulf oil both in direction and in magnitude. The new structure of fob differentials between the main oil exporting countries provided an "effective basis for the 'orderly marketing' of the greater part of the companies' incremental production of Middle Eastern crude."[96] The control mechanism of the companies which materialized in the 1950s as a result of recognized joint interests was in many ways comparable to the anti-competitive principles of the Achnacarry Agreement.

The cornerstone of their control was the almost complete domination of oil production outside the US. The nature of the concessions gave them a virtual monopoly over oil production in a country and enabled them collectively to exert their power through the supply of crude oil. Concessions in the Middle East were exclusive, like the concessions of Aramco in Saudi Arabia and IPC in Iraq. Development of the concessions needed to be complemented by extending the channels to process and distribute the oil. Since the large international oil companies faced structural changes in the main consumer markets after 1945, notably the US market, their market regulation power was increasingly based on their holdings in the traditional oil producing countries. This output of oil was managed through the consortia and long-term oil contracts by the international oil companies. Their grip on the industry was temporarily strengthened by agreements on market shares and cross-hauling between companies.[97]

The increase in posted oil prices in the 1950s, the return on these foreign investments, and the declining competiveness of American production drew new companies into international exploration. The expansion of demand, especially in Western Europe and Japan, allowed new supply to be brought on the market without a major decrease in prices. Nevertheless prices

[96] Ibid., p.56.

[97] In practice the 'As Is' Agreement and subsequent agreements survived the official abolishment after the US Senate had disclosed the existence of these agreements. Anti-trust suits were started but never finished. The companies were able to continue their practices of collusion abroad.

were declining gradually, and the control of the Seven Sisters was waning. The virtual closure of the American market for imports affected the short-term ability of the market to absorb supply. Prices fell, and competition between companies increased throughout the 1960s. This took place in a period when the price elasticity of demand was declining, and the substitution of coal by oil had already advanced.

The penetration of oil in the OECD economies had already induced the govenments to intervene in the domestic oil markets. Some govenments protected their local energy producing industries from competition from oil; other govenments favoured crude oil imports over oil product imports. Also the governments of the producing countries tried to increase their control over the industry. They tried to (re)negotiate concessions, royalty payments, taxation of company profits, and wanted to increase their control over production and pricing. In the 1960s, the oil companies, large and small, found themselves increasingly in the middle of a contest between producer and consumer governments, which were trying to cream off their economic rents.

Chapter Five

The Organization of the Petroleum Exporting Countries

1. The founding of the Organization of the Petroleum Exporting Countries

The Organization of the Petroleum Exporting Countries (OPEC) was established in September 1960 in Baghdad by Venezuela, Iran, Iraq, Kuwait and Saudi Arabia.[1] The establishment of the Organisation was a countervailing action on the part of the oil producing countries against the reductions of posted prices by the international oil companies.[2] The price reduction had taken place without prior consultation with the oil producing countries. Previous separate attempts by the governments of these countries to expand their influence on the production and pricing of oil had failed to be effective.[3] Since the Bandung Conference, the developing countries had discovered the potency of collective action, and they were pressing for international economic reforms in the United Nations. In 1960, the first development decade was announced and the the first UNCTAD conference was on the drawing board. For the oil producing countries, the price reductions were the straw that broke the camel's back.

The disputes between the governments and the oil companies again centered on the oil revenues of the governments. Since the revenues were determined by royalties and taxes per barrel and on posted prices, the income was mainly dependent on the volume of oil the companies could absorb into their integrated operations. The income of the producing countries was at disparity with the growth of world oil demand and their needs to develop their own economies.

[1] The Treaty to establish OPEC was published on September 24, 1960 after the inaugural meeting of the founding member states on September 10-14, 1960. After ratification by the member states, OPEC was registered as an international organisation at the United Nations on November 6, 1962. OPEC was officially recognized as an international organisation by the United Nations Economic and Social Council on June 30, 1965. J. Evans, op.cit., p.129.

[2] See A. Sampson, op.cit., 1985 and A. Alnasrawi, op.cit., 1985.

[3] Venezuela, Iraq and Iran had their clashes with the oil companies during the 1950s. For further reading see for example: P. Terzian, op.cit., 1985; A. Hussein, *Iraq: The eternal fire*, Third World Centre, London, 1981.

The Statute of OPEC[4] comprises 42 Articles, divided into VI Chapters. Chapter I determines the Organization and its objectives, and in Chapter II the conditions of membership of the Organization are laid down. Chapter III deals with the Organs of the Organization and their authority. Chapter IV considers Consultative meetings and Specialized Organs; Chapter V, the financial provisions, and Chapter VI the additional provisions.

The statutes reflect the priority of the founding member states on (stabilized) oil revenues. In Chapter I, Article 2, sub A to C, the foundation of the Organization's policy is formulated: Article 2, sub A: "The principal aim of the Organization shall be the co-ordination and unification of the petroleum policies of member countries and the determination of the best means for safeguarding their interests, individually and collectively."; sub B: "The Organization shall devise ways and means of ensuring the stabilization of prices in international oil markets with a view to eliminating harmful and unnecessary fluctuations."; sub C: "Due regard shall be given at all times to the interests of the producing nations and to the necessity of securing a steady income to the producing countries; an efficient, economic and regular supply of petroleum to consuming Nations; and a fair return on their capital to those investing in the petroleum industry."[5]

Resolutions IV 32-34 (4th Conference in Geneva 1962) elaborated on earlier statements with regard to prices and revenues, and on the principles laid down in the Statute.[6] General implementation of royalty expensing became OPEC's main issue in the initial negotiations with the oil companies. The negotiations were difficult. Saudi Arabia and Iran conducted the negotiations on behalf of the oil exporting countries. The companies were still crediting royalties in the Middle East, which had been a major concern to Venezuela, where royalties were expensed. The companies would not (or could not because of antitrust legislation[7])

[4] The OPEC is an intergovernemental organization and has three organs: The Conference, The Board of Governors, and the Secretariat. The Conference, the supreme authority of the Organization, has two ordinary meetings a year and consists of delegations representing a member country.

[5] The statutes of OPEC were approved at the Organzation's Conference in Caracas in 1961 by the adoption of Resolution II.5. The Statute was amended several times (at the 3rd, 4th, 5th and 6th meeting, fully revised at the 8th meeting in 1965 and amended by the 12th, 20th, 24th, 50th, 51st, and 57th meeting). The current edition of the Statute is dated June 1986. *OPEC Statute* June 1986 and *OPEC Official Resolution and Press releases 1960-1983*, Pergamon Press, 1984, Oxford, and the *1984-1987 Supplement*, OPEC, Vienna, 1987.

[6] Resolution IV 32 called for the restoration of crude oil prices to the pre-August 1960 level. Resolution IV 33 insisted that royalty payments should not be credited against income tax and should be set at a uniform and equitable rate. Resolution IV 34 attempted to eliminate any contribution of governments to the selling expenses of the companies. *OPEC Official Resolutions and Press Releases 1960-1983*, Oxford, 1984, p.16-18.

[7] "Of course, in many foreign nations there was no real comprehension of United States antitrust laws. Thus, it is unlikely that most OPEC leaders ever realized that, if the American companies joined together in negotiating production quotas and prices with OPEC, the Justice Department could charge them with violating antitrust laws." B.H. Wall, *Growth in a changing Environment; a history of Standard Oil Company (New Jersey), Exxon Corporation 1950-1975*, McGraw-Hill Book Company, New York, 1988, p.605.

recognize OPEC as a collective negotiating organization.[8] Whatever the legalities, it served the companies' purpose of stalling the negotiations on the royalty expensing. At the seventh OPEC Conference in November 1964 in Djakarta, the OPEC countries were in dispute over the latest offer of the companies. The offer encompassed royalty expensing, while the companies would get a diminishing allowance ($8^1/_2$% in 1964, $7^1/_2$% in 1965 and $6^1/_2$% in 1966) of the posted price, and could thus reduce tax payments to the producing countries.[9] Iran had concluded a tentative agreement with the companies before the Conference. Saudi Arabia, Kuwait, Qatar and Libya were also prepared to accept this offer, while Iraq vehemently opposed the proposal (supported by Venezuela and Indonesia, although these countries were not directly involved in this issue). The dispute illustrated the problems involved in acting collectively in a situation where interests were not entirely comparable, and where the international oil companies were in a strong position. The international oil companies negotiated through their consortia and could thus ensure collective bargaining while trying to circumvent anti-trust laws.

The companies not only faced pressure from the OPEC with regard to royalty expensing in the 1960s, but also from individual member countries. The Shah of Iran was pressing the Iranian Consortium to increase oil production more than the Middle East average yearly increase.[10] He wanted to increase his oil income to finance the economic and military ambitions of Iran. The oil market in the early 1960s was still relaxed, however, and extra production would have to be compensated by reduced growth of production elsewhere in the Middle East. The companies, with their balanced approach of production and consumption in the eastern and western hemisphere markets could not infringe on this balance. Beside the economic difficulties of this request, the political balance of the Middle East was just as important. The ARAMCO partners also needed to passify similar demands made by the Saudi government.

Meanwhile, the companies needed a positive rate of substitution for oil to ensure growing demand. The international oil companies had, however, to cope with increased competition from the Independents in their eastern hemisphere markets and increasing exports from the Soviet Union. The American market, where the Independents had more extensive processing operations, was at first virtually closed for oil from the Middle East and North Africa, forcing companies from these regions to develop new outlets for their oil on the European market.

Although the collective bargaining position of the large international oil companies towards OPEC was still strong in the early years of the Organization, this position weakened when

[8] Saudi Arabia and Iran negotiated on behalf of the exporting countries because the major companies were either party to ARAMCO or the Iranian Consortium, or both, thus skirting the legal objections of the oil companies.

[9] B.H. Wall, op.cit., p.608.

[10] In 1966 the Middle East average would mean a 10-11% increase of Iranian production. The Shah was pressing for an increase of 17.5%. B.H. Wall, op.cit., p.612.

competition was augmented from other oil companies and OPEC increased its membership from 5 to 10 states.

During the remainder of the 1960s OPEC mainly devoted attention to the development of petroleum policy. The 'Declaratory statement on petroleum policy in member countries' of June 1968 called for development of petroleum resources in the member countries, the possibility of government participation in existing operations, regulation of posted or tax reference prices and the establishment of price differentials according to gravity, quality and geographic location and sound conservation practices by the operators in the oil industry in the member countries. With regard to pricing, the 'Declaratory statement on petroleum policy in the member states' stated: "All contracts shall require that the assessment of the operator's income, and its taxes or any other payments to the state, be based on a posted or tax-reference price for the hydrocarbons produced under the contract. Such price shall be determined by the government and shall move in such a manner as to prevent any deterioration in its relationship to the prices of manufactured goods traded internationally. However, such a process shall be consistent, subject to differences in gravity, quality and geographic location, with the levels of posted or tax-reference prices generally prevailing for hydrocarbons in other OPEC countries and accepted by them as a basis for tax payments."[11]

2. Policy-making in the 1970s

When OPEC took over the determination of oil pricing from the international oil companies in 1973, it attempted to realize several policy objectives. Its main objective was to maintain oil prices in concordance with the development of prices of manufactured goods, and with prices of investment goods. The increased income was to serve the development needs of the economies, and any deterioration of their terms of trade might jeopardize these goals. This development dimension is important for the understanding of the market behaviour of the OPEC member states after 1974.

The market conditions of the early 1970s had brought about a process of concentration and facilitated OPEC to exert cartel power. In the early 1970s, price elasticity of demand and the short term supply elasticity were very small. Spare production capacity had diminished from the late 1960s onward, and the market was tight. Moreover, the OPEC member states were the most important net-exporters. Countries such as Libya, Nigeria and Algeria, where the companies had invested in the 1950s and 1960s, had joined OPEC. The decreasing oil prices in the 1960s had depressed exploration in more marginal areas, and OPEC oil was the only source of incremental supply. Only after the price increases of 1973/1974 was exploration stepped up. However, this oil only came on supply after 1978 in sufficient quantities to substitute for part of OPEC oil. Between 1974 and 1978, the market conditions seemed to

[11] Declaratory statement on petroleum policy in the member states, Resolution XVI.90, Vienna, June 1968, *OPEC Official Resolutions and Press Releases 1960-1983*.

remain favourable to the OPEC cartel, and the price fluctuations were all upward.

The notion that OPEC's market behaviour should be considered similar to a single unified market participant has led to many deceptive interpretations of its conduct, and consequently of the market behaviour of the other market particpants. The individual governments of OPEC countries have: "differing 'rational' economic interests depending on domestic social pressures, revenue needs, alternative sources of export earnings and fiscal income, hard currency financial assets, and geological reserves. Hence, they have different discount rates for present versus future earnings and different strains or pains associated with holding spare capacity or not developing additional capacity. Ultimately, the members of OPEC have different preferred price and production paths for the exploitation of their petroleum reserves."[12]

It is therefore not surprising that the OPEC member states, having gained autonomy over their national raw material sources for the first time, mainly pursued a policy of national self-interest after 1974. The member states were predominantly occupied to assume direct control over crude production and marketing and to replace the international oil companies control by government control. In the political and economic context of the countries, the decision to tranfer authority over price and production to OPEC was unimaginable. Moreover, OPEC is an intergovernmental organization, and decisions are based on consensus. As long as the member states share (enough) common interests, the Organization is able to represent the member states' joint decisions. However, this holds true for all collusive or cartel-like market behaviour. But bearing in mind that the member states had to pursue their energy policies within the broader context of their national economic policy goals, not to mention their diverging political interests, common interest-sharing among the national states is at variance with their national interests.[13] The fundamental assymmetry in objectives and stategies, and the fact that none of the member states or groups of member states could assert a strong political and economic leadership, made OPEC into an unstable group that could not effectively monopolise the market.

The unique short-term market conditions for OPEC, and the political circumstances in the Middle East in the early 1970s, had allowed them to reach a Nash equilibrium, but this could, as it turned out, not be repeated. However, these circumstances transformed the structure of the industry. The success of 1973 led the member states to believe that they could translate their strong bargaining position into power to regulate the market. However, such a role required that OPEC: (1) maintained the largest possible degree of symmetry or had to accept a strong leadership role of one or a group of member states; (2) had the authority to monitor the member' states supply; and (3) would incorporate the groups' environment in its decisions. The intergovernmental nature of the OPEC assured that the individual member states maintained their sovereignty over their oil policies, and that any

[12] T.H. Moran, op.cit., in: *International Organization*, Spring 1981, Volume 35, no. 2, p.244.

[13] See section 4, Chapter one, for a discussion of OPEC and the stability of the group.

agreement had to be arrived at through bargaining. These agreements either had to serve individual member states' interests, or should be sanctioned by the leader. Any agreement that failed to satisfy these prerequisites would invite cheating and increase the internal instability. This instability prevented the member states from facing their environment sufficiently into account, because internal agreements were already so hard to reach that external conditions could not be effectively dealt with. For example, their price policy did not sufficiently take the long-term elasticity of supply and demand into account, nor the market behaviour of outside players. OPEC was therefore destined to be more re-active than pro-active.

The immediate worry of the OPEC member states after the price increases of 1973/1974 was to safeguard the purchasing power of a barrel of crude. The determination of the OPEC member states to maintain real oil prices was again stressed[14] in a statement in 1975, when the inflationary policies of the industrialized consuming countries had already affected real oil prices in 1974 and 1975 (See Figure 3.2). The 'Solemn Declaration' of March 1975 reiterates some former opinions on prices and adds a linkage to prices of final oil products: "The sovereign heads of state point out that an artificially low price for petroleum in the past has prompted over-exploitation of this limited and depletable resource and that continuation of such policy would have proved to be disastrous from the point of view of conservation and world economy. They consider that the interest of the OPEC member countries, as well as the rest of the world, would require that the oil price, being the fundamental element in the national income of the member countries, should be determined taking into account the following: (i) the imperatives of the conservation of petroleum, including its depletion and increasing scarcity in the future; (ii) the value of oil in terms of its non-energy uses; and (iii) the conditions of availability, utilization and the cost of alternative sources of energy. Moreover, the price of petroleum must be maintained by linking it to certain objective criteria, including the price of manufactured goods and technology for the development of OPEC member countries."[15]

Notwithstanding the upward moving trend of nominal oil prices, the OPEC member states failed to maintain the purchasing power of oil with regard to OECD export prices. The oil producing countries were facing stagnating demand after the first oil price increase, and had to maintain their output at a fairly stable level (See Figure 3.19). Due to inflation, and depreciation of the dollar, their purchasing power decreased. The deterioration in real oil prices on the one hand, and decreasing demand in 1974 and 1975 (See Table 3.1) resulted

[14] Real oil prices had also deteriorated between 1970 and 1973, after the Teheran and Tripoli Agreements. The member states had in fact been negotiating with the oil companies to adjust oil prices to the level of inflation in 1973. The oil companies did not want a complete correction of oil prices and negotiations were broken off. When they were about to start again, the October War had broken out and OAPEC had announced its boycott and production cuts.

[15] Solemn Declaration, Conference of the Heads of State, Algiers, March 1975, *OPEC Official Resolution and Press Releases 1960-1983*.

in a decrease of the value of OPEC oil exports of 10.5% (See also Figure 5.1). In 1978, after two subsequent years of increasing value of oil exports, the value of oil exports decreased again by 4.5%. Between 1974 and 1978, the value of oil exports had increased 12.5%[16], but the value of OPEC imports cif had increased 192%.

The accumulation of income after 1974 diminished the incentive for the member states to activate a solid, unified, production and pricing policy. OPEC merely served as a discussion platform, where the main policy objectives were formulated, and as an information channel on other member states' policy decisions. Issues like pricing, output regulation, price diffentials and long-term strategy were still in 'study' in special committees. The Conference of OPEC was mostly dominated by immediate political problems that were a legacy from the 1973/1974 price increase and monetary matters.

In 1973, the OPEC established a price structure of Official crude prices, with Arabian Light as a marker crude.[17] The price structure was based on types and gravities of the OPEC crudes and on their geographic location. Except for the level of oil prices, the system was similar to the oil companies' system. Light crudes were priced higher than heavier crudes, and sweet crudes were priced higher than sour crudes, reflecting the fact that the yield of a barrel of refined heavy crude is lower than the yield of a barrel of light crude[18]. The differentials[19] were still an expression of the traditional supply and demand conditions of the crudes. The fact that the chemistry of supply was progressively dominated by heavier crudes, since additions to reserves in the 1970s were mainly additions to heavy crude reserves[20], and that the member states wished to reserve lighter crudes for domestic refining purposes, should have resulted in adaptions in price differentials.[21] Although the member states agreed in principle that the differentials should express the economic value of crude in terms of final products, should encourage a switch in demand to heavier crudes, and

[16] Between 1972 and 1974, the value of oil exports had increased 411.7%.

[17] A proposal of representatives of the OPEC countries' national oil companies in London in August 1974, regarding state oil-sales also included a proposal involving the observance of a loose system of market-demand prorationing, and of minimum selling prices equivalent to 93% of the posting for Arabian Light 34^0 crude fob Ras Tanura after appropriate adjustment for gravity, quality and geographic location was made. This proposal was not accepted by the majority of OPEC member states governments. However, the official pricing formula recommendation received general acceptance on an informal basis. J. Evans, op.cit., p.412.

[18] In the refinery crude oil is basically broken down into three basic groupings of products. The components of these three groups of products vary according to the gravity of the crude. A barrel of light (average API 44^0) breaks down into: 30% of residue (with fuel oil as an important product), 40% of middle distillates (with gas oil as an important product and naphtha) and 30% of light products (for instance gas, gasoline, aviation fuels, naphtha). Naphtha is extracted from both the light and middle distillate cuts. A barrel of crude API 31^0, still considered a light crude, can be broken down into: 50% of residue, 31% of middle distillates and 19% of light products. OPEC, *Basic Oil Industry Information*, 1983, p.28-29.

[19] Technically, the differentials equalize the refiners cost for diverse grades, sulphur and wax contents, and transportation charges over the yearly demand cycle. T.H. Moran, op.cit., in: *International Organization*, Autumn 1987, volume 41, no. 4, p.600.

[20] *OPEC Bulletin*, March 1982, p.1.

[21] See Chapter 1.

should facilitate fair competition between the crudes of each member state, the price structure was not adapted accordingly. Moreover, proposals to adapt the differentials were shelved several times, and the member states merely acknowledged the general principals of the pricing system. The decisions on pricing taken between 1974 and 1978 only concerned the maintenance of the purchasing power of a barrel of crude. Corrections for inflation and the dollar depreciation resulted in upward adjustments of the nominal oil prices, but basically left the pricing system intact.

Price differentials among crudes are, however, important and have progressively become a major point of dissent among the member states. Price differentials between crudes can stimulate or stifle demand for certain types of crude. The differentials can be used as a means of (unfair) competition among the member states, when the member states want to dodge the general pricing agreements. Price differentials, on top of barter agreements, extended credit, absorption of freight costs and net back pricing, were increasingly used by the member states for price shaving to increase or defend market shares.[22]

The differentials issue has been on the agenda several times but, except for some minor changes, it was never really resolved. Its very complexity, and the possible consequences for inter-OPEC competition among crudes, made the differentials a controversial subject. Changes in differentials can also influence the value of oil exports and the income of the member states. The fact that the member states principally pursued their national interests, lacked diversification in the downstream industry, and basically remained dependent on crude oil exports, made changes to new market conditions complicated. Acknowledging traditional differentials, and making changes on an *ad hoc* basis only, prevented OPEC from homogenising oil more efficiently. The few changes in differentials were mostly the result of plain threats and political pressure.[23] Since 1974, the internal difficulties in establishing a system of fair price differentials have only become larger. Market conditions changed, which diminished OPEC's ability to control the market. OPEC lost a substantive part of its market share to NOPEC producers, and consequently the issue of price differentials was no longer an exclusive OPEC issue anymore.

After 1979, competition between crudes increased. The break-down of the traditional marketing channels increased the volume of oil purchased on the spot market[24], increasing the number of short-term contracts. The restructuring of the market refineries enhanced input flexibility in terms of types of oil, and further encouraged competition between crudes based

[22] T.H. Moran, ibid., in: *International Organization*, Autumn 1987, volume 41, no. 4, p.600.

[23] Saudi Arabia, Kuwait, Qatar and the UAE were discontented in 1983 with prevailing price differentials between the value differentials of light African crudes to the Gulf crudes and with the discounts on official selling prices. The acceptance of new quota allotments was made conditional to a change in differentials and to an abolishment of discounts. In 1985, changes in differentials were accepted by the Conference but Algeria, Libya and Iran dissociated from this decision. Seventy-Third (Extraordinary Meeting of the Conference), Geneva, press release 3/85.

[24] See Chapter six.

on price differentials. NOPEC crudes were priced 'competitively' in relation to OPEC prices, and OPEC lost effective control over pricing. Individual member states were challenged in particular markets by NOPEC crudes, and some member states were forced to adjust their prices to the spot market prices. From 1982 to 1986, both real and nominal oil prices weakened, despite OPEC's output restrictions.[25] In December 1985, the OPEC pricing system collapsed altogether.

At the 80th Meeting of the Conference (December 1986) the official selling prices of the member states crudes were determined at approximately $18 per barrel. In conjunction with a re-established production policy, OPEC tried to get a grip on pricing again to bring the spot-market prices of OPEC crudes into line with the agreed levels of official reference prices. These price levels were not reached until April 1989, and since then the spot-market prices have once more declined below the official prices.[26] OPEC established in this meeting of the Conference a fixed pricing system with an $18 per barrel OPEC reference price. This price is based on a basket of seven crudes. At the Conference, the problems of values of OPEC crudes relative to the reference price was discussed with a view to achieve fair competition between crudes. To guarantee this, the Conference decided that the maximum value difference between Arab Heavy (Saudi Arabia) and Bonny Light (Nigeria) should not exceed $2.65 per barrel.[27] However, the price differential between these two crudes on the spot-market had reached $3.14 in June 1989.

Another important factor in discussions on price relations between crudes of the member states are the differentials with competing NOPEC crudes. In the first quarter of 1989 the Middle Eastern benchmark, Dubai crude, was $2.50-$3.00 lower than North Sea Brent, a doubling of the differential since 1988. The stronger demand for light refined products like petrol increased demand for Brent, which yields more light products than Dubai crude.[28] In June 1989, the differential between these crudes had diminished to $2.15.

In order to homogenise oil prices effectively, the determination of fair price differentials must take several factors into account: a) the quality of the crudes; b) the chemistry of supply; c) the chemistry of demand; d) transport costs. Development of an efficient OPEC

[25] See section 4.

[26] In June 1989, OPEC spot-market prices were in the range of $14.71 for Saudi Heavy and $15.40 for Dubai crude to $17.85 for Bonny Light and Saharan Blend. The spot-market prices for NOPEC crudes were in June 1989 (4 week average): Oman Blend $ 15.80; Suez Mix (Egypt) $15.11; Brent $17.55; Ekofisk $17.78; Isthmus (Mexico) $18.30; West Texas Intermediate $ 20.04; Urals (Soviet Union) $16.61. *OPEC Bulletin*, July/August 1989, p.29-32.

[27] At the 73rd meeting of the Conference, January 30 1985, price differentials were also discussed, and the Conference decided that the maximum differential between heavy crude produced in the Gulf and light crudes produced in Africa should around $2.40 and in accordance with: Arab Heavy $26.50, Arab Medium $27.40, Arab Light $28.00, Murban $28.15, Minas $28.53, Bonny Light 28.65, as of 1 February 1985. Press Release 3/85 Geneva, 30 January 1985 and 8/86, Geneva 20 December 1986, in: *OPEC Official Resolutions and Press Releases 1984-1988*, p.228 and *OPEC Bulletin*, December/January 1987, p.3.

[28] *The Economist*, April 8, 1989.

pricing system will remain an arduous task as long as the member states approach this issue in terms of national economic and political interests only, continuing to disregard the market conditions. Table 5.1 illustrates the price differentials of official selling prices with respect to the OPEC reference price of $18 per barrel.

Table 5.1 Price differentials among OPEC crudes

country	crude	API°	$ price
Algeria	Saharan Blend	44.0	18.87
	Zarzaitine	42.0	18.87
Ecuador	Oriente	29.5	17.02
Gabon	Mandji	30.5	17.32
Indonesia	Minas	34.1	17.56
Iran	Iranian Light	34.0	17.50
	Iranian Heavy	31.0	16.89
Iraq	Basrah*	35.0	17.30
Kuwait	Kuwait Export	31.0	16.67
Libya	Brega	40.0	18.67
	Zueitina	40.5	18.67
Nigeria	Bonny Light	36.7	18.92
Qatar	Dukhan	40.0	17.82
	Marine	36.0	17.67
Saudi Arabia	Berri	39.0	17.87
	SA Light Ras	40.0	17.52
	Tanura	31.0	16.92
	SA Medium	27.0	16.27
	SA Heavy		
UAE	Murban	39.0	17.92
	Umn Shaif	37.0	17.72
	Zakum	40.0	17.82
	Dubai Fateh	32.0	17.42
Venezuela	Tia Juana**	24.0	13.27

*:not agreed upon at the 80th Conference in December 1986, but is the official Iraqi selling price; **:Venezuela has minimum sales prices and, since 1986, prices for some crudes are contract or market-related prices, no official price was agreed upon at the 80th Conference, price quoted is price December 17, 1987; source: OPEC Annual Statistical Bulletin 1987, Table 71.

Sovereign governments acting as market participants are governed by more incentives than to collectively optimise prices and price differentials alone. The differences between the member states in terms of economic development, absorbtion capacity, oil output and reserves, and to a certain extent the political and social system of the countries resulted in widely diverging required oil price targets. The efficiency of the OPEC in optimizing pricing and production relies on the ability and willingness of the member states to yield national policy goals to collective policy goals. With regard to pricing and price differentials, OPEC has not (yet) been very efficient, and has not been able, considering the once favourable market conditions for the cartel, to maintain control over pricing.

Although many features of the oil companies' pricing system were copied by the OPEC official pricing system, an important feature was left out. The companies established a two-basing point system and, after 1948, equalized oil prices in New York. The equalization of prices enabled the oil companies to service their main consumer markets with crudes from their international production. The cross-hauling between companies reduced costs of transportation, which compensated for the sales of less-competitive crudes. Moreover, the companies priced crude within the context of both their upstream and downstream activities. Needless to say, some of the advantages of the companies' pricing system were negated by import-controls of the American government and (price) competition of the Independents, emphasising once more the limitation in time of the efectiveness of cartels.

The lack of downstream processing facilities and distribution networks in the main consumer countries affected OPEC pricing. Oil income was related to crude exports and prices only, and not to income derived from final oil products. Although OPEC pricing was supposed to take (long-term) price elasticity of demand, changing chemistry of demand, and inter-fuel competition into account, these issues disappeared into the background at the end of the 1970s. The prevalence of short-term over long-term gains contradicted earlier formulated policy goals. The insistence on maintaining a high price level in a weak market required such output constraints that OPEC's market control evaporated. OPEC pricing has not been very efficient in terms of securing cartel stability. The high oil prices invited many new entrants to the industry and stimulated conservation and substition away from oil to an undesired degree. Moreover, the deterioration of cohesion between the member states accelerated after 1979.

3. The cohesion of OPEC

The cohesion of OPEC was considered to be guarded by the provision of Article 7 of the Statute. New members must fulfill the conditions of membership as laid down in Article 7, sub C of Chapter II of the Statute. These conditions are: "Any other country with a substantial net export of crude petroleum, which has fundamentally similar interests to those of member countries, may become a full member of the Organization, if accepted by a majority of three-fourths of full members, including the concurrent vote of all founding members."[29]

The phrasing of this Article is important. Potential member states should be net exporters of crude petroleum and should have **fundamentally** similar interests to the existing member

[29] *OPEC Statute*, June 1986, p.5.

states of OPEC. The admission of Qatar in 1961, Indonesia and Libya in 1962, Abu Dhabi in 1967[30], Algeria in 1969 and Nigeria in 1971 concurred with these conditions, at least at the time of their admission. However, the admission of Gabon (associate member in 1973 and full member since 1975) and Ecuador (associate member in 1973 and admitted as a full member at the end of 1973) is less obvious and seemed more inspired by political reasons than because of their 'fundamentally' similar interests in oil production. The extent to which the member states shared common interests, however, and were prepared to act upon them, was decisive for the market power the Organization could exert.

3.1 Dependency on oil exports

In reality, the economic and political differences among the member states were already substantial at OPEC's inception, and these increased over time. The diverging pace of economic development, the level of oil production, the proven reserves, the growth in domestic demand, the diversification of the economy, and the social and political structure of the member states, changed the face of OPEC and its power to regulate the international oil market.

Already in 1965, the member countries differed widely in terms of their oil production capacity[31] and in terms of the share of oil exports in total exports. The share of oil exports in total exports in 1965 of Venezuela, Saudi Arabia, the United Arab Emirates, Qatar and Libya was more than 90%. The share of oil exports in total exports in Iran was slightly lower, 87.2% and was 77.3% for Algeria. The shares of Kuwait, Indonesia, Nigeria, Gabon and Ecuador were substantially lower, and were respectively, 58.7%, 38.3%, 25.4%, 12.1% and 1.3%.[32] After the oil price increases the share of oil exports in total imports increased for all member states, except for Libya[33], and decreased on average only after 1983. In 1988 the shares of oil exports in total exports were for the member states: Algeria 67.4%, Ecuador 44.5%, Gabon 66.6%, Indonesia 26.7%, Iran 89.9%, Iraq 99.1%, Kuwait 87.9%, Libya 96.2%, Nigeria 93.9%, Qatar 91.4%, Saudi Arabia 86.4%, United Arab Emirates 60.0% and Venezuela 84.8%.[34]

In terms of dependency on oil exports, the similarity in interest of the member states increased in the formative years of the Organization, but proved to be a major destabilizing

[30] In 1974 Abu Dhabi's membership was transferred to the United Arab Emirates.

[31] In 1965 Algeria produced 558,700 b/d, Ecuador 7,800 b/d, Gabon 24,900 b/d, Indonesia 480,600 b/d, Iran 1,908,300 b/d, Iraq 1,312,600 b/d, Kuwait 2,360,300 b/d, Libya 1,218,800 b/d, Nigeria 274,200 b/d, Qatar 232,600 b/d, Saudi Arabia 2,205,300 b/d, United Arab Emirates 282,200 b/d and Venezuela 3,472,900 b/d. *OPEC Annual Statistical Bulletin 1985*, p.44-55.

[32] *OPEC Annual Statistical Bulletin 1985*, p.8.

[33] Libya depended 100% on oil imports in 1965, and this share in total exports decreased slightly after 1965 but remained high in the 90-100% range.

[34] *OPEC Annual Statistical Bulletin 1988*, Table 7, p.8.

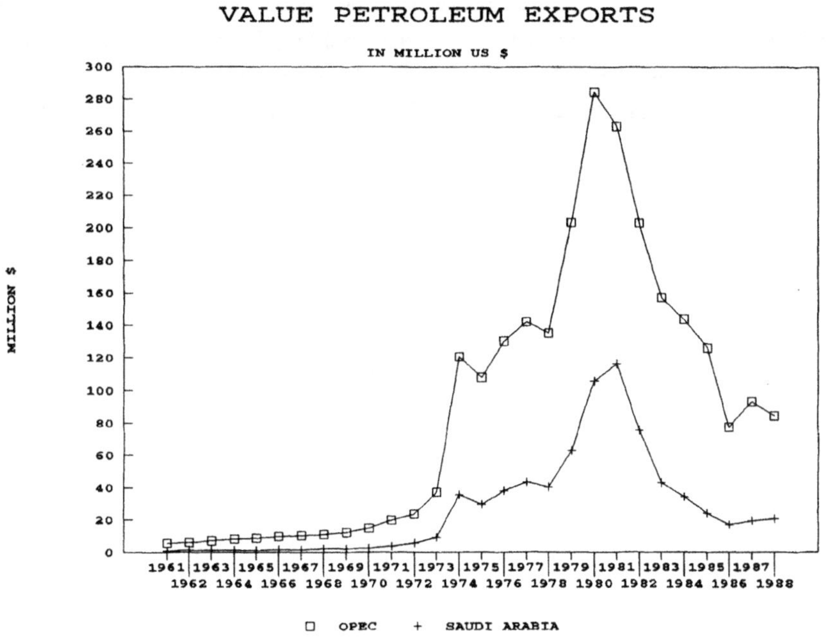

Figure 5.1

factor to cohesion after 1980, when the OPEC introduced the production quota.[35] The loss of export-earning capacity as a result of production cuts was substantial (See Figure 5.1), and created serious current account deficits for several member states.[36] The divergence in economic disequilibria between the member states in the 1970s and 1980s structurally altered the national (energy) policy goals of the individual member states. This seriously affected the stability of the cartel.

In 1970, the value of the OPEC member states' petroleum exports (including oil products) was $14,501 million.[37] In 1973, these earnings had increased to $37,164 million and to

[35] As a result of the decision to stabilize prices at $34 per barrel OPEC was required to cut production to 19 million b/d, a decline of 15.9% from 1981 to 1982. The brunt of the reduction in output was taken by Saudi Arabia. Saudi Arabia was producing almost at near full capacity in 1980 and 1981. In 1982 output was reduced by 33.9%. *OPEC Annual Report 1982*, p.14.

[36] The loss of income was a result of a changed price elasticity of demand in the OECD countries, the competion from non-OPEC oil (mainly from the North Sea, Alaska, Mexico and the USSR) and substitution of fuel oil for coal. The ability to accomodate to a loss of income appears to be larger in low-absorbing countries. However, comparing the dependency on oil exports to total exports, the low-absorbers have less opportunity to implement demand switching policies and current accounts deteriorated quickly.

[37] *OPEC Annual Statistical Bulletin 1987*, Table 5, p.6.

$120,457 million in 1974, an increase of 224% in 1974/1973. In 1980, the earnings had increased to $286,984 million. From 1980 onwards, the value of petroleum export-earnings started to decline, to $206,655 million in 1981 and to $130,681 million in 1985. The oil price collapse in 1986 caused a large decline in the value of oil export-earnings. In 1986, the value of OPEC export-earning declined to $78,834 million. In 1987, export-earnings

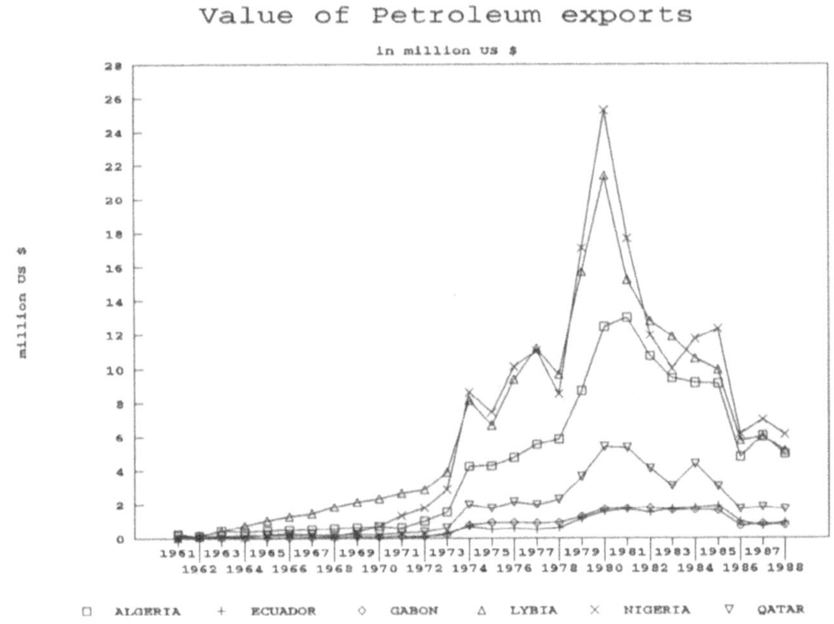

Figure 5.2

recuperated somewhat to $95,699 million.

The impact on the value of petroleum export-earnings of the two oil price increases (1973/1974 and 1979/1980) and the subsequent collapse in 1986 varied among the individual member states (See Figures 5.1, 5.2 and 5.3). The development of the value of export-earnings of Kuwait, Indonesia and Venezuela is fairly similar, and related to price and output developments. In the development of export-earnings of Iran, the impact of the loss of production as a result of the revolution in 1979 and the war with Iraq is obvious. Iraq's loss of earning coincides with the beginning of the war with Iran in 1980. The magnitude of the increase and subsequent decrease in earnings of Saudi Arabia is immense. The decision of Saudi Arabia to serve as the OPEC's swing-producer from 1981 onwards has resulted in a sharp decline of export-earnings. It is save to say that OPEC's pricing and production policy failed to fulfil the Organization's objective to stabilize income from oil. The large

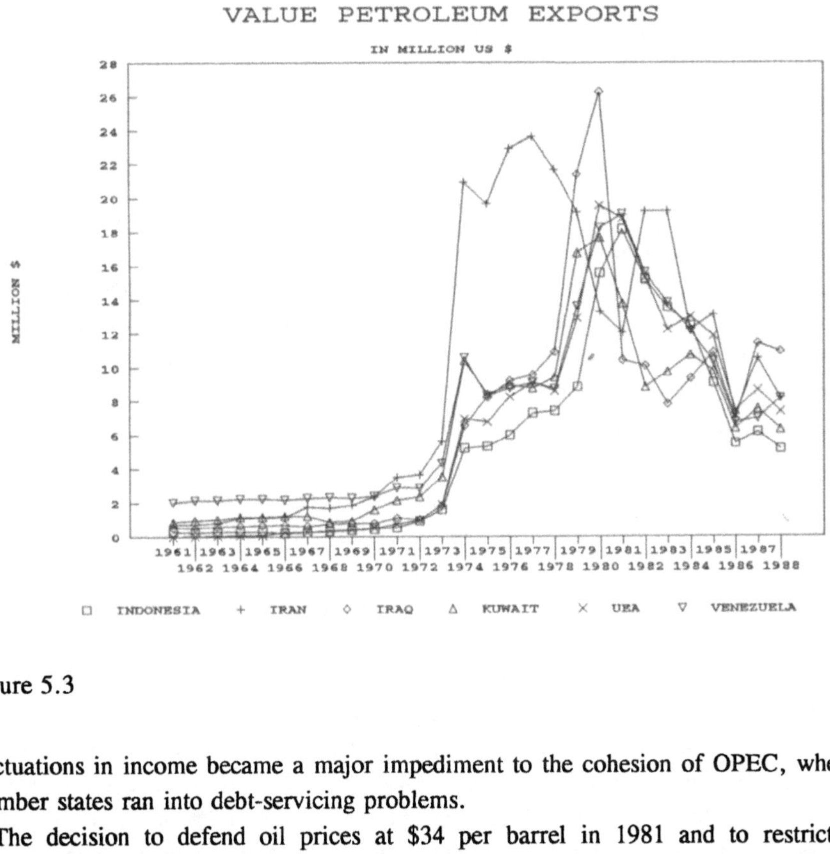

Figure 5.3

fluctuations in income became a major impediment to the cohesion of OPEC, when some member states ran into debt-servicing problems.

The decision to defend oil prices at $34 per barrel in 1981 and to restrict OPEC production obviously had severe consequences for OPEC's market share. From 1981 onwards, OPEC was a quantity-adaptor and a price-taker. In 1975, OPEC's market share in world oil production was 51.5%, and 71.8% of world exports. In 1978 these shares had decreased slightly to 49.4% and 69.6% respectively. In 1981 its share in world oil production had already decreased to 40.4%, deteriorating quickly thereafter. In 1983, the share had dropped to 32.3%, declining to 30.6% in 1984. After the price collapse in 1985/1986, OPEC's share increased slightly to 32.6%. OPEC experienced a similar downward trend in its share of world crude oil exports.[38] Not surprisingly, the member states experienced peak production levels between 1977 and 1979 (except for Ecuador who

[38] In 1981 this share was 58.6%, in 1983 46.8% in 1984 45.2%, in 1985 and in 1986 46.5%. *OPEC Facts and Figures*, 1983 and 1987.

experienced its peak production level of 280,000 b/d in 1985).[39]

3.2 *Proven reserves and production capacity*

The allocations of the output restrictions among the member states were a further strain on the stability of the cartel. The larger producers were forced to cut production disproportionally to their capacity and proven reserves. The logic for these disproportionate cuts was the dependency on oil income of the high absorbers, combined with the relatively small possible cuts in production in smaller producing countries.

The disparities among the member states in terms of proven reserves and the r/p ratio are large (See Table 5.2). The enormous reserves of Saudi Arabia, the United Arab Emirates and Kuwait, with a combined share of 44.6% of world oil reserves in 1990, and 58.2% of OPEC proven reserves, sets them apart in OPEC. The balance of power in OPEC has increasingly shifted to these Gulf countries. Moreover, they consult each other on policy matters and establish the broad lines of OPEC policy among themselves. The reserves of Iran and Iraq are also impressive, but political and economic differences have prevented these countries from aligning themselves with the other Gulf countries, nor with any other member state. The pattern of proven reserves in the Gulf region was not only a source of regional political and economic tension, but also provoked global political and military intervention when Iraq occupied Kuwait in August 1990.

Compared to other major oil producing countries like the USSR, the US and Mexico, the reserves of the Gulf countries countries were immense.[40] The proven reserves of the USSR were 57.0 thousand million barrels in 1988, 5.6% of the world proven oil reserves. The US reserves were 33.9 thousand million barrels, 3.4% of world proven oil reserves, and the proven reserves of Mexico were 52.0 thousand million barrels, 5.2% of the world proven oil reserves. The proven oil reserves in United Kingdom and Norway were respectively 3.8 (0.4%) and 7.6 thousand million barrels (0.8%). The size of the reserves of the Gulf countries, and the size production capacity in relation to their domestic consumption, made them assume the role of market leader in OPEC.

The r/p ratio of Saudi Arabia, Iraq, Kuwait and the United Arab Emirates exceeded more than 100 years in 1990. In terms of the r/p ratio, Venezuela compared favourably to the Gulf countries with a ratio of 71.0. Just as in Saudi Arabia, Iran and Iraq, recent oil discoveries

[39] Peak production levels of the member states: Algeria 1,161,200 b/d in 1978, Gabon 223,000 b/d in 1975, Indonesia 1,686,200 b/d in 1977, Iran 6,021,600 b/d in 1974, Iraq 3,476,900 b/d in 1979, Kuwait 3,283,000 b/d in 1972, Libya 3,318,000 b/d in 1970, Nigeria 2,302,000 b/d in 1979, Qatar 570,300 b/d in 1973, Saudi Arabia 9,900,500 b/d in 1980, United Arab Emirates 1,998,700 b/d in 1977 and Venezuela 3,708,000 b/d in 1970. *OPEC Annual Statistical Bulletin 1987*, p.44-53.

[40] *BPSR 1989*, p.2.

had led to a major upward appraisal of reserves in Venezuela.[41]

The other member states had either more moderate to small reserves and/or moderate r/p ratios, and did not compare to Saudi Arabia, Kuwait, Iran, Iraq and the United Arab Emirates. This disparity between the member states revealed a fundamental conflict of interest with respect to short- and long-term policy goals. Member states with both large proven reserves, a high r/p ratio, and a low absorption capacity (Saudi Arabia, Kuwait, United Arab Emirates, and Libya), benefit from policies that ensure a low rate of substitution. This implied a relatively low oil price (compared to alternative resources). The necessary high production levels do not immediately jeopardize their future position in the oil industry. The economic value of oil in the ground is lower than for the other member states or other oil producers. The rate of substition, the energy coefficient, and the oil/GNP ratio of OECD countries[42] suggested that the major oil exporting member states should at least anticipate a period of stagnant or slower growing demand in the major consuming countries, and consider this seriously in their policy decisions. A world 'hooked' on oil secures their income from oil for the future.

Member states with smaller reserves, a moderate production capacity, a relatively low r/p ratio, and a high absorption capacity (Gabon, Ecuador, Algeria, Nigeria, and Indonesia) gained from a relative high oil price. The value of oil in the ground is larger for them than for the first group of member states.

A last category of member states (Iran, Iraq and Venezuela) share common interest with both other groups. They were characterized by large reserves, a large production capacity, and a high r/p ratio, but also a high absorption capacity. At the end of the 1970s, they, particularly Iran and Iraq, aligned with the group that gained from a relatively high oil price, thus serving their short-term interest only. As a possible pressure group within OPEC, these two countries have lost some importance. The conflicts between Iraq and Iran, and with the other member states, alienated them from the mainstream in OPEC. Venezuela has in recent years aligned itself with the other Gulf producers, and Iran has recently taken a more cooperative position. This left Iraq in an isolated position.

[41] The r/p ratios of the Soviet Union, the United States, the United Kingdom, Norway and Mexico are respectively: 13.6, 10.3, 5.6, 12.2, and 49.9 years. *BPSR 1990*, p.2.

[42] Already in the OPEC Annual Report of 1978 attention is drawn to the fact that the average energy coefficient of the OECD countries was estimated by the IEA to be around 0.83 for the period 1977-1985, "showing a downward trend from the historical 1.0-1.1 period". "This, however, reflects only a global picture because there is a wide range of differences among the countries, due to the fact that the energy elasticity for any country will depend on a variety of factors concerned with the structure of the economy, the energy mix employed, the efficiency of energy use, and the population structure. However, over a period of time, through technological progress, conservation and energy efficiency, energy elasticity will be reduced." *OPEC Annual Report 1978*, p.57. *See* for the mentioned developments in the EC, R. Lefeber and J.G. van der Linde, op.cit..

Table 5.2 OPEC proven reserves, percentage share of world reserves, r/p ratio and production at the end of 1990

country	reserves in thousand million barrels	% share	r/p**	production thousand b/d 1989	production thousand b/d 1990*
Venezuela	59.0	5.8	71.0	2,010	2,365
Iran	92.9	9.2	81.6	2,845	3,125
Iraq	100.0	9.9	>100	2,825	2,005
Kuwait	94.5	9.4	>100	1,640	1,065
Saudi Arabia	257.5	25.5	>100	5,280	6,700
Ecuador	1.4	0.1	13.9	295	280
UAE	98.1	9.7	>100	2,035	2,300
Qatar	4.5	0.4	27.8	410	455
Algeria	9.2	0.9	21.8	1,205	1,275
Libya	22.8	2.3	46.5	1,160	1,350
Nigeria	17.1	1.7	27.1	1,635	1,810
Indonesia	11.1	1.1	20.5	1,415	1,520
Gabon	0.7	0.1	8.0	215	250
OPEC	773.8	76.6	87.2	23,340	24,775

*: production levels per country were affected by the Gulf crisis, i.e. the boycott of Kuwaiti and Iraqi exports, and the subsequent increase in production by other OPEC memebr states; **:r/p reserves at end of year divided by production of that year, result is number of years of production at the level of that year's production; source: BPSR 1989, and for Gabon Annual Statistical Bulletin 1988.

3.3 *Absorption capacity*

The absorption capacity of the member states became a major factor in decision-making and defining short- and long-term (energy) policy goals. Absorption relates to the capacity to invest capital productively. The possibilities of enlarging the absorption capacity of the economy are limited in extent and speed. Absorption is defined as: "The amount of investment that can be made at some threshold rate of return, with the supply of complementary factors considered as given."[43]

The absorption capacity is determined by a wide range of factors such as: the structure of the population, the structure of the economy, the labor market (potential), schooling and the political structure. The OPEC member states varied widely with regard to population and the

[43] D.J. Teece, OPEC Behavior: An Alternative View, in: J.M. Griffin and D.J. Teece (eds.), *OPEC Behavior and World Oil Prices*, 1982, p.67.

size of the economies. The significance of the oil industry as a welfare generating sector to the economies of the member states, differed qualitatively and quantitatively.

An indication of differences among the member states can be illustrated by the differences in GNP per capita (See Table 5.3). Indonesia, Nigeria and Ecuador rank among the lower middle-income countries, while Algeria, Gabon, Iran, Iraq and Venezuela rank among the higher middle-income countries.[44] Kuwait, Libya, Qatar, Saudi Arabia and the United Arab Emirates rank among the high-income oil exporters, and compare well with the GNP per capita of the highest ranking industrial market economies, like the US, Norway and Switzerland. After 1981, Saudi Arabia experienced a reduction of GNP per capita compared to the higher levels of GNP per capita in the 1970s.[45] The GNP at market prices of Saudi Arabia, UAE, Kuwait, Libya, Nigeria, Qatar, Indonesia and Venezuela decreased sharply since 1981.[46] Ecuador and Gabon experienced more 'moderate' decreases, while Iran and Iraq accomplished increases. The dependency on oil exports evidently relates positively to the development of GNP at current market prices.

The ability to diversify the economy and to increase the absorption capacity of the Gulf states is also correlated to the labour market potential. The qualitative and quantitative limitations of the labour markets in the member states creates "a decline in the marginal productivity of capital resulting from the inability to augment the human factors of production as fast as the capital stock."[47] Member states with a limited labour force[48], like Saudi Arabia, Kuwait, Qatar, UAE have finite investment possibilities. The increase in export revenues spurred demand for imported consumption and capital goods and services. Domestic production could only satisfy expanded demand to a very limited extent.[49] However, the scarcity of labour restricts the possibility of developing a diversified domestic industry. Increasing, for instance, the production capacity of the relatively labour intensive consumer goods sector, would unnecessarily strain their scarce resources.[50] Kuwait has, for this

[44] *World Bank Development Report 1988*, p.222-223.

[45] In 1985 GNP per capita was $7,916 and in 1986 $6,413. In 1985 the population of Saudi Arabia was 11,520,000, in 1986 11,650,000 and in 1987 13,610,000. The increase of the population of 2 million (1987 over 1986) may partly explain the reduction. However, the large increase in population (18%) seems unrealistic and the increase must be due to new estimates by the government. *OPEC Annual Statistical Bulletin 1985 and 1986.*

[46] See Table 3, *OPEC Annual Statistical Bulletin 1987.*

[47] D.J. Teece, op.cit., in: J.M. Griffin and D. J. Teece (eds.), *OPEC Behavior and World Oil Prices*, 1982, p.67.

[48] The labour force is also restricted, because women are hardly participating in the labour process. Islamic law does play a role, although in other Islamic countries a larger part of the female population is actively involved in the labour market.

[49] Y.A. Sayigh, Arab Economic Strategy in a Changing World Oil Market, in: *Third World Quarterly*, January 1982, p.18.

[50] Of course the level of technology in industry could improve the capital/labour ratio. Demand for labour could be qualitative instead of quantitative. However, training of local labour takes a long time. On top of that, the allocation of the national labour force has been directed at 'white collar' labour in the private service sector and public sector of the economies. The lack of low and medium trained technicians inhibits industrial developments.

reason, strongly developed its foreign investment portfolio and yielded in 1987 an estimated $6.3 billion of income compared to $5.4 billion of income from oil exports.[51] Investment in export oriented capital and energy intensive industries also has its limitations. It makes the economy very dependent on world market conditions, technology transfer, and the small labour force was not available for these sectors. Local labour was mainly productive in the traditional sectors of the economy, while skilled nationals were progressively employed in the service and public sector of the economy.[52] The countries have tried to ease the tension on the labour market by importing labour on a massive scale. However, wages of skilled foreign labourers are high. Migrant un- and semi-skilled labour was mainly employed in the industrial sector and construction sector. The social structure of these societies, however, and the dependency on foreign labour in key functions in the private and public sector of the economy induced the governments to change the labour market policies.[53] Saudi Arabia introduced limitations to foreign labour imports, stringent rules regarding the duration of work-permits, and stricter guidelines for remittances. These restrictions were coupled to a 'Saudiisation program' to reverse the dependency on migrant labour, especially in the modern and public sector of the economy.[54]

OPEC member states with large populations were faced with the opposite problem. The absorptive capacity of the economy of these countries outstripped their income from oil. The development of their economy demanded enormous capital commitments in infrastructure, agriculture, industry, services and human capital. With respect to capital commitments necessary to develop the infrastructure, they do not differ from the low-absorbing member states. However, the investment requirements in the industrial sector, agriculture and services were many times larger. Domestic industry could not meet the increase in demand for consumption, capital goods, and services, and was mainly met by imports. Between 1973 and 1980, prices of imports from the industrialized countries increased 225%, due to price and

Foreign labour almost exclusively holds these positions in industry.

[51] *The Economist*, March 26, 1988, p.59.

[52] C. van der Linde, *International Division of Labor and Economic Integration of the Arab World*, University of Amsterdam, March 1984, unpublished.

[53] Ibid., p.35. The member states of the Gulf Co-operation Council (GCC), established in 1981 to co-ordinate among other things their industrialization plans, try to develop their economies through efficient use of scarce production factors. They have established joint-ventures in dock-yards, aluminum smelting and in steelplants.

[54] "Economic pressures arising from the fall in export revenues have underlined the urgency of Saudiisation at all levels of government, industry and commerce. The seriousness with which the authorities are approaching this issue, reflected in a sustained high level of budget allocations for education and training at a time of stringency in most other sectors, suggests the kingdom's dependence on imported labour will fall steadily over the next five to 10 years. However, the need for rapid qualitative as well as quantitative improvements means that, in the short term at least, there are tremendous opportunities for specialist foreign companies and agenies to become involved in the training effort. The 1987 budget allocation for manpower resource development is SR 23,725 million ($6,325 million), a cut of only 1 per cent on 1985/86, and roughly 22 per cent of the total SR 109,248 million ($29,130 million) allocated to civilian expenditure." *MEED Special Report Saudi Arabia*, June 1987, p.53.

cost inflation.[55] The current accounts of Algeria, Ecuador, Indonesia and Nigeria have been in deficit most of the 1970s and 1980s.

Table 5.3 GNP per Capita in US$ of OPEC member states in 1987, 1988 and 1989

	1987	1988	1989
Algeria	2,689.7	2,284.7	1,929.2
Ecuador	933.4	1,016.1	910.8
Gabon	2,288.2	2,400.0	2,559.8
Indonesia	372.2	393.3	477.3
Iran	3,432.6	3,408.2	3,604.5
Iraq	3,145.9	3,021.9	3,716.8
Kuwait	14,449.2	13,159.7	15,199.5
Libya	4,639.2	4,349.9	4,485.2
Nigeria	317.4	242.4	163.7
Qatar	15,305.7	15,645.7	17,845.5
Saudi Arabia	5,398.9	5,534.9	5,462.2
UAE	16,191.6	15,527.3	17,669.0
Venezuela	1,885.2	3,202.4	2,137.4

source: OPEC Annual Statistical Bulletin 1987, 1988, and 1989 p. 2 and 3.

The average annual growth rates of gross domestic investments between 1965 and 1980 were impressive in all member states. However, this rate substantially slowed down or became negative in the period 1980-1986. This was an indication of the economic problems the member states encountered as a result of the reduction of oil income.[56] Some member states experienced serious debt servicing problems in the 1980s. The current account balances of some member states have been negative (Indonesia, Algeria, Ecuador, Nigeria) for extended periods between 1967 and 1987.[57] To finance spending and consumption growth, these

[55] M.M. El-Imam, Decline in Arab Oil Revenues, in: A.M. Farid and H. Sirriyeh (ed.), *The Decline of Arab Oil Revenues*, Croom Helm, London, 1986, p.21.

[56] The average annual growth rate of Gross Domestic Investments was between 1965-1980 and 1980-1986 respectively: Indonesia 16.1% and 3.7%; Nigeria 14.7 % and -13.7%; Ecuador 9.5% and -5.6%; Algeria 15.9% and 0%; Venezuela 8.4% and -8.4%; Gabon 2.4% for the period 1980-1986; Iran 11.6% for the period 1965-1980; Iraq not specified; Saudi Arabia 27.5% for the period 1965-1980 and not specified for 1980-1986; Kuwait 11.7% for the period 1965-1980 and not specified for 1980-1986; UEA not specified; Qatar not specified and Lybia 7.3% for the period 1965-1980 and not specified for the period 1980-1986. *World Bank Development Report 1988*, Table 4.

[57] See *OPEC Annual Statistical Bulletin 1988*. Also Saudi Arabia experienced a current account deficit since 1983.

member states had borrowed heavily from the international banks with oil as the prime collateral. The loans were concluded with the expectation that oil prices would continue to rise. When prices started to decrease, at first in nominal terms, and after the Plaza agreement of 1985 also in dollar terms, the oil producing countries were in serious payments trouble.[58] The total external debts[59] in 1986 of the indebted member states were: Indonesia $42,090 million; Nigeria $21,876 million; Ecuador $8,953 million; Algeria $17,929 million; Venezuela $33,891 million; and Gabon $1,568 million. In 1987, Iraq was reported to have accumulated a debt of $60 billion to foreign lenders.[60] No information on Iranian debts was available, but the war efforts had also induced them to engage in substantial borrowing abroad. The other OPEC member states had no outstanding debts. The total long term debt service as a percentage of GNP and the exports of goods and services in 1986, disclosed the extent of exposure of the member states. Debt servicings as a percentage share of GNP and exports were respectively: Indonesia 7.3% and 33.1%; Nigeria 3.3% and 23.4%; Ecuador 8.3% and 33.9%; Algeria 8.7% and 54.8%; Venezuela 8.6% and 37.4% and Gabon 7.1% and 17.5%.[61] It suffices to remark that establishing production quotas among the member states with 6 out of 13 members in payment difficulties, and Iran and Iraq at war until August 1988, were the main causes for the disproportionality of quota allocations to production capacity. The continuance of this situation led to the price collapse of 1986 and cheating on the quota by various member states.

3.4 Downstream development

The varying rate of growth of downstream developments of the member states also diminished the shared interests of the member states. Following the price increases of 1973/1974, the member states expanded their downstream activities. The percentage share of OPEC in world production of refined products (excluding CPEs) increased from 8.7% in 1970 to 13% in 1989 (See Table 5.4). Consumption of refined products also increased, from 1.9% of world consumption (excluding CPEs) in 1970 to 7.4% in 1989. OPEC production of refined products increased between 1970 and 1989 with 3,041.6 thousand barrels a day, an increase of 94.5%. OPEC consumption of refined products increased in the same period with 2,867.8 thousand barrels a day, an increase of 420.5%.

In 1989, Saudi Arabia produced 23.6% of OPEC's refined products, Venezuela produced 16.1%, Kuwait 12.5%, Indonesia 10.2%, Iran 10.4%, Algeria 7.5% and the other member

[58] Real oil prices had already deteriorated since 1982, See Figure 2.1. On top of the deteriorating purchasing power of a barrel of oil and the subsequent payment difficulties, interest rates on the oustanding loans were high.

[59] Total external debt: Long-term debt (public, publicly guaranteed and private non-guaranteed), short-term debt, use of IMF credit, short-term debt. *World Bank Development Report 1988*, Table 16.

[60] The GCC countries are reported to have borrowed Iraq $30 billion. *Financial Times*, 2-9-1987, p.3.

[61] *World Bank Devlopment Report 1988*, Table 19.

states the remainder. OPEC's share in world exports of refined products was 29.5% and 34% of NCW exports. The share of Saudi Arabia in total OPEC exports of refined products was 29.8% in 1989, that of Kuwait 19.2%, that of Venezuela 17.2% and that of Algeria 10.9%.[62] The predominance of this group of OPEC member states in the production and exports of refined products created another (potential) conflict of interest. The development of downstream operations, and the subsequent investments of some state oil companies in downstream operations in the major OECD markets, created a change in the national policy objectives of these member states. The growth of refined product exports became an incentive for some member states to produce more oil than their OPEC output quota prescribed. In 1987 Kuwait's oil exports consisted of almost 90% of refined oil products, and the country managed to sell 20% more than its OPEC output quota of 1 million barrels a day.[63]

Table 5.4 OPEC Refinery production, consumption and exports in thousand b/d

year	capacity	production	consumption	share of OPEC prod. in NCW
1970	3,565	3,217	682	8.7
1975	4,488	3,042	1,281	7.4
1980	6,237.4	3,983.8	2,398.6	8.2
1985	6,429.5	5,003.6	3,199.2	11.9
1989	6,926.4	6,258.6	3,549.1	13

OPEC Statistical Bulletin 1975, 1981, 1989.

Another incentive for member states with substantial petrochemical facilities to press for extended production quota, and accept a lower crude oil price, was the link between oil production and the produced volume of associated gas. Associated gas was increasingly used as a fuel and feedstock in the refining and petrochemical industry, thus linking the output of crude oil to the output of the processing industry. The enormous investments in the processing industry, some in joint-venture with foreign (oil) companies, required a certain capacity utilization rate to get a fair return on the invested capital. The substantial crude oil production cuts of Saudi Arabia in the beginning of the 1980s reduced the available feedstock for new processing. In 1984 and 1985, Saudi Arabia produced below its allocated production quota, and the capacity utilization rate of its processing industry neared the minimum

[62] Exports include re-exports and exports of petroleum products from gas plants. *OPEC Annual Statistical Bulletin 1989*, Table 27.

[63] *The Economist*, March 19, 1988, p.59 and October 15, 1988, p.93.

requirment to ensure a fair return on their investments in the downstream industry.[64] Saudi Arabia's role of swing-producer came at a cost. In the Autumn of 1985, Saudi Arabia abandoned this role, increased production, and implemented net-back pricing, to re-establish its position in the market.

3.5 Political friction among the member states

On top of economic differences between the member states, several political and military intra-OPEC disputes occured in the period under review. In June 1961 Iraq invaded Kuwait (and tried again in 1973 and indeed occupied Kuwait in August 1990) immediately after Kuwait's independence, but withdrew when Britain supported Kuwait.[65] Part of the border is still disputed. The Rumaila oil field stretches across this disputed area. Iraq never gave up its claim on the Kuwaiti island Bubiyan. The claim on this island was revived after the conclusion of the Iran-Iraq war, when Iraq developed two alternative ports on the Khawr Abd-Allah, the waterway it shares with Kuwait, to its ports on the Shatt al Arab. Iraq has only a limited coastline and shares its access to the Gulf with either Iran and Kuwait. Iraq's occupation of Kuwait in August 1990 certainly had political and strategic origins, but the growing dispute with Kuwait and other member states over oil policy brought the conflicts to a climax.

Iraq and Iran had several border disputes over the Shatt-el-Arab, and have also supported the Kurdish resistance to undermine each other's governments. In 1975, an agreement was reached to settle the border disputes, and promises were made to refrain from supporting the Kurdish resistance. The border settlement was again disputed, however, and was now made part of the peace-negotiations following the Iran-Iraq war (1980-1988).[66]

Iran and Saudi Arabia have had tense political relations, both during the Shah regime and after the proclamation of the Islamic republic. The struggle for regional power and religious supremacy in Islam has influenced their relations continuously. The bloodshed in Mecca during the Hadj in 1987 resulted in a further deterioration in the relations of the countries.

Iran's occupation of several islands in the Arabian/Persian Gulf is still a source of discontent among Iran and the UAE, and Qatar. These small OPEC member states also have border disputes among themselves.[67]

In November 1988, Saudi Arabia and Kuwait settled their dispute over oil-rights in the

[64] Friedrich Ebert Stiftung, Internationale Politik, *Saudi-Arabien in den 80er Jahren*, Expertengespräch 26. und 27. April 1982, p.21-22.

[65] Iraq claimed Kuwait as part of the country based on the Ottoman empire.

[66] Iraq claims the agreement was reached under pressure and therefore not valid. The peace negotiations are, among other reasons, deadlocked because of it.

[67] Qatar has brought the dispute over the reef Fasht al-Dibl with Bahrain (a non-OPEC member state but a member state of OAPEC and the GCC) to the International Court of Justice in the Hague. *The Economist*, January 21, 1989.

Neutral Zone.[68] A border dispute between Saudi Arabia and the UEA dating back to 1949 was only settled in 1974.

Another dispute in the Gulf region could well erupt out of the Iranian threat to tap parts of Arab Gulf offshore reserves. Gholamreza Aghazadeh, the Oil Minister of Iran, declared that seismographic tests had revealed that several oil and gas fields already exploited by the southern Gulf States extend into Iran's continental shelf. The Minister said he saw no legal reason why Iran should not exploit the joint fields off Qatar and the UAE. Two possible joint fields off the coast of Saudi Arabia are still under study.[69]

4. Conflicts on pricing and output

The conflicts of interest among the member states became explicit in 1979, when OPEC had to deal with the sudden temporary termination of Iranian oil production. The Iranian revolution in 1979 and the temporary termination of oil production had created panic among some oil consuming countries. Although Saudi Arabia increased production by 1,231.5 thousand b/d in 1979/1978 and Iraq by 914.9 thousand b/d, which more than compensated for the loss of Irani production of 2,073.8 thousand b/d, the oil market was in turmoil. Companies and oil consuming countries were, with the crisis of 1973/1974 only a few years past, competing for oil to build-up stocks.[70] Public stock-piling in the OECD countries was in many countries handled by the (private) oil companies. The stocks of the companies can, therefore, be divided into required stocks and discretionary stocks.[71] The companies' management of discretionary stocks, and other consumers, is not conducive to countering excessive price movements: "Thus, expectations of dearth and of rising prices prompt them to build inventories; on the other hand, expectations of surplus and of declining prices prompt them to liquidate inventories. It was acknowledged that the companies' stockpiling scramble of 1979-80 exacerbated price increases, while the reverse process accelerated price

[68] The countries did, however, share production in the Neutral Zone for a long time. *The Economist*, January 21, 1989.

[69] *Financial Times*, 17-3-1989.

[70] Yamani commented on the behaviour of companies and oil consuming countries with regard to stocking: "As we all know, stocks play an important role for any commodity: when there is a surplus, you buy for your stock to replenish it; and when there is a shortage, you draw from your stock. But what has been happening in the oil industry is exactly the opposite. When we had a shortage because of the Iranian Revolution in 1979, companies started to buy heavily to stock up; and when we had a surplus in 1981 and 1982, those same companies started drawing heavily from their inventories. That is of course not a healthy situation, but it is a fact." Sheikh Ahmed Zaki Yamani, Statement of Saudi Policy (1982), in: R. Mabro (ed.), *OPEC and the World Oil Market; The genesis of the 1986 price crisis*, Oxford Institute for Energy Studies/Oxford University Press, 1986, p.69.

[71] Required stocks represent the minimum operating levels for a certain period of time, imposed for 'security reasons' by the governments or international organizations (i.e., the EC, the OECD or the national governments), and they are linked to current consumption or imports. In the OECD and EC, these stocks were increased during the crisis from 60 days of consumption to 90 days of consumption. Discretionary stocks are company stocks and the level is decided upon by the companies.

declines in 1982-84."[72] In 1979, OPEC called repeatedly on the large international oil and trading companies to refrain from speculating on the spot market, and to stop reaping windfall profits.[73] Initial attempts by OPEC to counter the upward pressure on the oil price by increasing production and restraining the official selling prices were quickly abandoned.

The situation on the oil market was aggravated when BP, Exxon and Socal (Chevron) declared a 'force majeure' on a large number of their long-term contract customers.[74] BP lost close to 1.4 million barrels per day, Exxon 400,000 thousand b/d, and Chevron 300,000 thousand b/d in long-term contracts. A number of refiners which had obtained crude through long-term contracts with these companies (the Third Party market), bought crude on the spot market or directly with the oil producers. Under the circumstances they were willing to pay a premium for crude. Consuming countries like Japan, which relied to a large extent on supplies from Iran, were also willing to pay a premium for crude. As a result, the system of long-term supply contracts, which passed the crude through the channels of the major oil companies, broke up, and was replaced by short-term contracts. The collapse of part of the Third Party market radically changed oil trade. More and more oil was traded on a short-term basis or through the spot-market.

Soon, the price developments on the spot-market induced some member states to divert oil to the spot-market, capturing higher prices than the official OPEC selling prices. This dispersion of official OPEC prices was condoned by OPEC. Already on 28 February 1979, a press release was issued, stressing the sovereignty of the member states to determine their oil price: "Conference decisions in setting crude oil prices do not prevent Member States from making an upward adjustment in the light of their prevailing circumstances. Therefore, the decisions taken by the Member Countries, either by adjusting their prices upwards, or by choosing other courses of action or further consultation, certainly fall within the framework of the Statute of the Organization, and are strictly in accordance with their sovereign rights, approved by the OPEC Conference as long ago as January 1961. The Conference emphasizes the "sovereign equality" of Member Countries and that "due regard shall be given at al times to the interests of the producing nations...". In the present

[72] Z. Mikdashi, *Transnational Oil, Issues, policies and perspectives*, Frances Pinter Publishers, London, 1986, p.31.

[73] OPEC consultative Meeting on 21 February 1979 in Geneva, Fifty-Third (Extraordinary) Meeting of the Conference in March 1979 in Geneva and in Press Release no. 7-79, Vienna, 18 September 1979, *OPEC Official Resolutions and Press Releases 1960-1983*, 1984. Yamani commented on the behaviour of the oil companies: "Throughout 1979, continued high demand in the west and the peak prices being paid on the spot markets were creating chaos. Oil companies were rushing in to buy. But this wasn't oil they needed. It was for storage. I asked myself why and the only possible answer was that they thought there could be a total cut-off in oil exports from Iran." In November 1979, American Embassy personnel were taken hostage in Teheran, adding to these fears. Another major factor to increased spot market buying was the dependence of Japan on Iranian exports. The revolution was considered a serious threat to the security of supply. J. Robinson, *Yamani, The Inside Story*, Simon & Schuster, London, 1988, p.214-215.

[74] Y. Stournaras, *Are oil price movements perverse? A critical explanation of oil price levels 1950-1985*, Oxford Institute for Energy Studies, WPM 6, 1985, p.44-45.

circumstances, the actions of Member Countries in exercising their sovereign rights cannot be construed as prejudicing the solidarity and unity of OPEC." [75] These 'sovereign rights' were price adjustments by individual member countries to spot market oil prices and to increased sales on the spot market, that is 'sovereign rights' to divorce themselves from the official policy of OPEC.

The former trading structure was disrupted when more member states funneled oil into the spot-market and engaged in direct contracts, by-passing the companies. The initial insecurity about the new trade structure, and the continued instability of Iranian oil production, increased the scramble for oil by traders, companies, and governments alike. Consequently, the oil price on the spot market soared to approximately $40 per barrel. The OPEC Secretariat repeatedly called for limitation of spot market transactions by all market participants. The February press release had already turned these calls into weak gestures, however, and emphasized the lack of power and authority of the Organization to discipline its member's market behaviour.

In the formative years of OPEC, the member states' common (external) interest (to gain control over price and production) over-shadowed their internal conflicts. In 1973, the combination of political action and favourable market conditions enabled the member states to establish sovereignty over their oil resources, and OPEC more or less stumbled into its existence as a cartel. After the euphoria had died down, the member states had to manage the oil market. Their external common interest had been achieved, and when policy had to be formulated, the internal conflicts became more apparent. Between 1974 and 1978, the member states were unable to agree on substantial policy matters, and the Iranian crisis prompted the fundamental internal conflicts to become explicit.

In 1980, the member states had to decide at what new level price should be stabilized. The smaller producing member states were in favour of a high official OPEC price, but especially Saudi Arabia was a strong proponent of more moderate oil prices.[76] However, the majority of the member states decided not to compensate for the loss of Iranian oil production any longer, instead reducing production, creating an even tighter market. In 1980, production of all member states, except Saudi Arabia, declined compared to 1979 production levels. OPEC production declined 4,011 thousand b/d from 1980/1979. Saudi Arabia continued to produce at full capacity in 1980 and continued to sell oil at lower prices than the agreed OPEC level. In June 1980, the official OPEC marker crude ceiling was established at $32, and value differentials were restricted to $5 above the marker crude price level.

In 1981, the majority of the member states again decided, with the exception of Saudi

[75] Press Release no. 3-79, Vienna, 28 february 1979, *OPEC Official Resolutions and Press Releases 1960-1983*, 1984, p.163.

[76] See Chapter 3.

Arabia[77], to cut production (by 10%) and to defend the new 'market' prices.[78] World consumption exceeded production by 475 thousand b/d in 1981, resulting in a serious deadlock within OPEC over Saudi Arabia's production policy. This breach of solidarity was strongly criticised. The policy of Saudi Arabia was explained by Yamani, as cited in Evans: "Speaking on US television on April 19, he said that 'we engineered the [oil]glut, and we want to see it in order to stabilize the price of oil', adding that the latest round of price increases by other OPEC members had 'gone too far' and was 'not in the interests of the international economy'."[79]

Saudi Arabia was unable to sustain its production at full capacity for a very long time and could not compensate for the decline in output of the other member states. Saudi Arabia lost its position as market leader and had to restore its cooperation with the other member states. Eventually the OPEC price structure was reunified in October 1981[80], and the official price for Arabian Light fob Ras Tanura was established at $34 per barrel. Saudi Arabia agreed to reduce its level of production[81] in 1982, restored in its role as OPEC's swing producer. Prices were determined in the range of $32-$36 a barrel, with Saudi Arabia remaining in the lower range and the rest of the member states in the higher range of prices.

The unification of pricing in OPEC was concluded in an already weakening oil market. World oil consumption decreased 1,695 thousand b/d from 1980 to 1981 (and 1,720 thousand b/d excluding CPEs) and from 1981 to 1982 world consumption decreased another 1,425 thousand b/d (and 1,460 thousand b/d excl. CPEs). The companies, expecting the oil price to come down, started to draw on their stocks. The companies' stocks had increased from 78 days of forward consumption in 1978 to 99 days in 1981, but in 1982 stocks had already declined to 94 days of forward consumption.[82] Destocking by the companies resulted at the end of 1981 and the beginning of 1982 in an upward adjustment of world supply with 2.5 million b/d, a volume that exceeded the exports of all oil-exporting countries, except for Saudi Arabia.[83] OPEC was obliged, in the course of 1982, to restrict output to compensate for the additional crude on the market.

The positive difference between official selling prices and spot market prices disappeared

[77] Saudi Arabia reached its peak production in 1980 and in 1981 production was only slightly lower-9,555,000 b/d. Late 1979, Yamani repeatedly stated that Saudi Arabia would not give in to pressure to restrict oil production. Saudi Arabia continued throughout the oil crisis to honour their long-term contracts and sold any additional oil at the OPEC market price. J. Robinson, op.cit., 1988, p.215.

[78] The marker crude was established at $36 and the maximum OPEC price at $41. Press Release no. 2-81, Geneva 26 May 1981.

[79] J. Evans, op.cit., p.567.

[80] Press Release no. 7-81, Geneva, 29 October 1981.

[81] See Figures 3.18 and 3.19.

[82] Z. Mikdashi, op.cit., p.31.

[83] G. Luciani, op.cit., p.67.

in 1982. In 1981, the average spot market price for Arabian Light[84] had been $37.65 (first quarter) against the official selling price of $32. In the third quarter of 1981, the spot market price was already below the official selling price ($31.95). The unification of OPEC pricing policy in October 1981, and the determination of the market crude (Arabian Light) at $34 a barrel relied on the success of the restriction of Saudi production. However, spot market prices (except in the fourth quarter of 1981) continued to decline. In 1982, the average sport market price (first quarter) declined to $31, $29.30 in 1983, $28.55 in 1984 and $27.78 in 1985. The official selling prices followed this downward trajectory with some delay, when the restriction on oil output failed to reverse the price trend.

The member states had obviously mistaken the spot market price, or the price of the 'panic barrel' for the market price, and on top of that, they had miscalculated the effect of the price increase on demand. Maintaining their official selling prices required substantial output restrictions, more than they thought they had bargained for.

In April 1982 the 63rd meeting of the OPEC conference established for the first time production quotas to restrict supply, and to maintain, on average, their official selling prices (See Table 3.3). The ceiling on oil production was placed at 17,500 thousand b/d in 1982 (See Table 5.5). Immediately after its conclusion, the agreement of 1982 was put to the test. Iran made its intentions clear to produce more than its allocated quota, and demanded an increase of of its quota of 500 to 700 thousand b/d at the expense of Saudi Arabia's quota.

The implementation of the production quota also worsened Nigeria's economic situation. Its allocated quota remained almost 1,000 thousand b/d under the level required to finance current government spending. Moreover, Nigeria had to maintain an official oil price well above competitive North Sea crudes. As a result, the country was forced to implement some emergency measures, including restricting imports.[85]

Factual OPEC production was eventually brought back from 23,380 thousand b/d in 1981 to 19,930 thousand b/d in 1982, and still remained 2,430 thousand b/d above the agreed ceiling. Prices were continuously under pressure. The spot oil prices were below official government prices, and according to the oil companies in January 1983, the net-back value for Arabian Light was more than $4 below the official government price.[86] Moreover, Saudi Arabia, Kuwait, Qatar and the UAE made the abolition of discounts and revised price differentials mandatory to their agreement with new quotas. In March 1983 at the 67th meeting of the Conference in London, new quota allotments were accepted, while maintaining projected total output at 17,500 thousand b/d.

Despite the maintenance of the OPEC ceiling and the shifts in the individual member states' quotas, production in 1983 was 18,425 thousand b/d, that is still 925 thousand b/d in excess of the quota agreement. In 1984, new quota agreements had resulted in a lower OPEC

[84] Z. Mikdashi, op.cit., p.35.

[85] J. Evans, op.cit., p.584/585 and See section 3.3 regarding Nigeria's debt servicing problems.

[86] Ibid., p.597.

ceiling of 16,000 thousand b/d. However, 1984 production even exceeded 1983 production (18,470 thousand b/d in 1984), due to over-production from various member states. Among the notorious over-producers or countries that offered discounts on the official government selling prices were in the period 1982-1985: Nigeria, UAE and Iraq[87]. After 1986 Iran, Kuwait and Saudi Arabia also occasionally exceeded their quotas. In 1983, prices were adjusted downward from $34 per barrel (Saudi Arabian Light fob Ras Tanura) to $30, and later to $29. In 1984 the price was maintained at $29, but had to be reduced to $28 in 1985. In 1986, prices were left to the market and decreased steeply on the spot market from $26.92 in December 1985 to $8 in July 1986.[88]

Table 5.5 OPEC production quota 1982-1991; in thousand b/d

country	April 1982	March 1983	Nov 1984	1986/1987 **	Jan 1989	Nov 1989	July 1990	2Q 1991
Algeria	650	725	663	700	695	827	827	827
Ecuador	200	200	183	200	230	273	273	273
Gabon	150	150	137	200	166	197	197	285
Indonesia	1,300	1,300	1,189	1,200	1,240	1,374	1,374	1,443
Iran	1,200	2,400	2,300	2,400	2,640	2,640	3,140	3,217
Iraq	1,200	1,200	1,200	1,500	2,640	2,640	3,140	0
Kuwait	650	1,050	900	1,000	1,037	1,037	1,500	0
Libya	750	1,100	990	1,000	1,037	1,037	1,233	1,425
Nigeria	1,300	1,300	1,300	1,300	1,355	1,355	1,611	1,840
Qatar	300	300	280	300	312	312	371	399
Saudi Arabia*	7,000	5,000	4,353	4,300	4,524	4,524	5,380	8,034
UAE***	1,000	1,100	950	900	988	988	1,500	2,320
Venezuela	1,500	1,675	1,555	1,600	1,636	1,636	1,945	2,235
Neutral Z*	300	-	-	-	-	-	-	-
Total	17,500	17,500	16,000	16,600	18,500	18,500	22,491	22,298

*since 1983 production of the Neutral Zone was included in quota of Saudi Arabia and Kuwait; they share production on a 50-50 basis; Saudi Arabia's quota was considered a maximum ceiling for its 'swing' production; ** between 1986 and 1988 interim quota arrangements were established, sometimes for two months only. These are approximately the levels of these agreements, while minor upward and downward adjustments occured;*** UAE did not accept its quotum for 1990; source: John Evans, op.cit., p.584, 601 and 624 for 1982, 1983, and 1984; and OPEC Official Resolutions and Press Releases 1984-1988 for 1984, 1986/1987 and 1989; *The Economist*, October 15, 1988 for 1986; *MEES*, 7 May 1990 for November 1989; Mees, 10 June for July 1990 and 2Q 1991.

[87] Iraq refused to agree to the quota system in 1983 and the subsequent agreements all have a 'theoretical' Iraqi quota allotment. Iraq demanded parity to Iran's share in production. This was eventually acknowledged in the 1989 agreement. *Financial Times*, 29-11-1989.

[88] *OPEC Bulletin*, September 1986, p.39.

When these quota allocations are compared to the production capacity (let alone proven reserves) of the member states, the irritation of some member states becomes readily understandable. The quota agreement of January 1989 meant that: in 1989 Algeria could only produce 65% of 1988 production; Ecuador, 76.7% of 1988 production; Gabon, 83% of estimated capacity and 97.6% of 1988 production; Indonesia, 72.9% of estimated capacity and 94.3% of 1988 production; Iran, 88% of estimated capacity[89] and 116.4% of 1988 production; Iraq, 97.8% of estimated capacity[90] and 101.5% of 1988 production; Kuwait, 51.6% of estimated capacity and 69.1% of 1988 production; Libya, 69.1% of estimated capacity and 76% of 1988 production; Nigeria, 75.3% of estimated capacity and 99.3% of 1988 production; Qatar, 62.4% of estimated capacity and 86.7% of 1988 production; Saudi Arabia, approximately 53% of estimated capacity[91] and 83.5% of 1988 production; the UAE, 45.5% of estimated capacity and 57.1% of 1988 production and Venezuela, 65.4% estimated capacity and 85% of 1988 production.[92] Collectively, OPEC produced at approximately 60% of total estimated capacity and 87.5% of its 1988 production level. Member states with a low quota compared to production capacity- Kuwait, Saudi Arabia and the UAE- have been bearing the brunt of the output restrictions since its commencement in 1982. Presently, Saudi Arabia insists on producing 24.45% of OPEC production, regardless of the production ceiling.

The cease-fire between Iraq and Iran caused the next strain on the production agreement. Both countries had announced large new reserves additions, which they wanted to develop, and with their post-war reconstruction under way, they soon started lobbying for larger quotas. Also other countries increased their production, which must be seen, in part, as strategic manoeuvring for the soon to be expected round of new production allocations. In April 1989, Iran was reported to have exceeded its quota by 10%, Kuwait by 35% and the UAE by almost 50%.[93] The policy of OPEC was geared to restore production regulation and to stabilize oil prices (on the spot market) at, or close to OPEC's reference price. However, achievement of the $18 per barrel reference price depends on the discipline of the member states and on co-operation from NOPEC producers. In the first quarter of 1990, OPEC was producing 1 million b/d above the agreed ceiling of 23 million b/d, and oil prices declined under the agreed reference price. The market showed signs of saturation, and the member states once again had to discuss their lack of discipline.

[89] Iran is expanding its production capacity as a result of post-war reconstruction, and is aiming at its pre-war production capacity of 6,600 thousand b/d rapidly. *Financial Times*, September 27, 1988.

[90] Iraq's production capacity was supposed to expand quickly. In 1989 Iraq was expected to increase the capacity to at least 3,600 thousand b/d or 4,000 thousand b/d. Under the present allocation scheme the share of production to capacity will decrease. *Financial Times*, September 27, 1988.

[91] *The Economist* (October 15, 1988) estimated the production capacity at 8,000 thousand b/d. However, Saudi Arabia produced close to 10,000 thousand b/d in 1980.

[92] Production capacity estimates from *The Economist*, October 15, 1988.

[93] *The Economist*, April 8, 1989.

These discussions took place under a cloud. Iraq had in the course of the first quarter of 1990 signalled that it was dissatisfied with certain member states' overproduction. Iraq particularly criticised Saudi Arabia, Kuwait and the United Arab Emirates. Until that time, Iraq had tacitly supported the low price policy of the Gulf states knowing that low oil prices would economically damage Iran more than Iraq. Moreover, after the cease-fire, Iraq endorsed the policy with the expectation that it soon would be producing significantly more than 3 million b/d, and wanted and needed market share for its oil. Early in 1990, Iraq completely reversed its market strategy, when the forecasts about its production capacity had proved to be too optimistic and debt-servicing became a serious problem. Iraq required increased revenue from oil exports to sustain its debt-servicing and to maintain investment levels in the oil industry. Already in 1989, Iraq had found it hard to find financial backing for its projects, and neither did flirting with the oil companies bring any of the desired investment capital. Instead, Iraq started to lobby for production restraints, and higher oil prices. Soon Iraq progressed from petitioning for the desired policy change to threatening neighbouring Kuwait, when policy adjustments seemed unforthcoming. First, Kuwait was accused of stealing Iraqi oil from the Rumaila field, and Iraq demanded reparations. Then it claimed Bubiyan island, and wanted its debt of about $10 billion with Kuwait annulled. Iraq stepped up its pressure on Kuwait when troops were moved close to the Kuwaiti border. In July 1990, after a period of flurried negotiations, the OPEC member states agreed to establish new quotas and to curb production until a price level of $21 a barrel was reached. This fell $4 short of the $25 that Iraq had demanded, but the general feeling was that a crisis had been averted. On 2 August 1990, Iraq invaded Kuwait, and the crisis immediately became international. The potential shift in market power when Kuwaiti and Iraqi production were to be combined, in addition to the regional political ambitions of Iraq's leader Saddam Hussein, provoked a strong reaction of the Gulf states, the US and West European countries. In January 1991, after a period of unproductive negotiations, forces of an anti-Iraq coalition, consisting of the US, some Western European countries, some Arab countries (among them Saudi Arabia, Egypt, Syria), and some other countries, took military action to drive Iraq out of Kuwait.

Shortly after Iraq's occupation of Kuwait, oil prices increased sharply because oil and product exports from Kuwait and Iraq were made subject to a United Nations boycott, which effectively took 4,5 million b/d off the market. Customers of Iraq and Kuwait saw their contracts forfeited and had to acquire oil on the spot market. The loss to the market of Kuwait's oil product exports was also serious. In certain regional markets, capacity was too short to compensate for this loss, and prices of particularly of gasoline were pushed up. However, the decision of the majority of the OPEC member states and other oil producing countries to increase production soon relaxed the crude oil market, and crude oil prices decreased. When the hostilities commenced on 17 January 1991, stocks were released by the IEA member states. This release of stocks combined with the coalition's immediate achievements, which removed the danger to Saudi Arabia's production, drove the prices quickly further down, to reach pre-August levels. In the aftermath of the war, prices remained stable, and OPEC faced the difficult task of accomodating its member states' new production objectives. Many countries had to invest to increase capacity in Fall 1990, and are now less inclined to return to much lower production levels. As long as Iraq's and Kuwait's production is in not on stream, or not allowed on the international market, the other member states can enjoy their higher production levels. However, the market is still too soft to accomodate these production levels when Kuwait and Iraq would resume production again. Also, both countries are in need of income to repair war damages.

While the member states have to tackle the short-term problem of a saturated market, they must also make provisions for an anticipated capacity shortage in the future. Since 1979, the sustainable OPEC production capacity has decreased from about 40 million b/d to 28 million b/d, and the 4.5 million b/d of unused OPEC capacity represents only 8% of world demand excluding the CPEs.[94] The member states want to increase their production capacity to sustain future oil production and meet the estimated growth in demand of 2% per year in the 1990s. There is no doubt that OPEC potentially has enough oil to meet growing demand. Some member states, however, foresee problems financing these additions to capacity. OPEC's Secretary General, Subruto, estimated the cost of a 20% increase of OPEC capacity at $60 billion. After the Gulf war, these costs have only increased. Although this estimate has initially been criticized as too high, the tight financial situation of some member states has brought about a new discussion on the level of the oil price. The emerging difference in pricing strategy again splits OPEC between the 'haves' and the 'have nots'. Some member states are increasingly courting the oil companies to invest in their capacity expansion, and new contracts, giving the companies equity and bigger profit margins on their investments, are already underway in Venezuela and Nigeria, and other countries are expected to follow suit.[95]

As much as the consumer governments have tried to realise security of supply, oil producing countries are now seeking security of demand. The producer governments have witnessed an erosion of their revenue, and they have seen their economic rents being taxed away by consumer govenments. The environmental policy measures, which include taxes on fuels, are bound to check demand in the main consuming countries. The producer governments stress that the investments necessary to meet future demand require some certainty about demand. New initiatives are taken to commence a dialogue between consumer and producer governments. However, is seems highly unlikely that these initiatives will bear any fruit. The interests between the parties do not converge, and governments, despite their increased intervention in the market, are not the only players in the market. The international oil companies still play an important role, and the overtures by some producing countries towards these companies to gain access to their cash flow and technology attest of their new

[94] *Petroleum Intelligence Weekly*, March 12, 1990, p.6.

[95] Ibid, 24 June 1991 and 1 July 1991.

market power.

5. Conclusion

The interests of the thirteen OPEC member states are politically distinct and economically divergent. The countries are party to the Statutes of the Organization of Petroleum Exporting Countries, an inter-governmental organization. In the formative years of the Organization, they shared a common (external) interest. They wanted to gain sovereignty over their own oil resources. The wish for political as well as economic self-determination characterized relations between developing countries and the industrialized countries in the 1960s and early 1970s. In those years, much attention was devoted within international fora to the behaviour of multinational companies and their (disputed) benefit to the economies of developing countries. The seven major oil companies were perceived as the examples par excellence. The fact that the oil producing countries were very dependent on their income from oil exports, and thus on the oil companies market behaviour, made the struggle for sovereignty over their resources a prime objective. When the member states asserted sovereignty over their oil resources it was perceived as being as much a political as an economic victory. However, after 1973 these states were not only in a position to decide on their domestic energy policies, but they also embarked on the arduous task of trying to manage the international oil market.

The achievement of their common (external) interest in 1973 required re-evaluation of their policy objectives and the translation of these objectives into practical market management. This task drew out the fundamental internal conflict of interest between the member states. The oil price crisis of 1979 caused a rift in the Organization. The majority of the member states opted for stabilizing the price at a high level. This price level was more in concordance with the price of a 'panic barrel' than a sustainable market price. The attempt to maintain the oil price required substantial output restrictions, especially as demand for oil slumped and NOPEC oil production increased. In the process, OPEC lost its dominant market position, and it gradually became the international oil market's 'swing-producer'.

The member states will have to face up to the reality of the market developments and their internal conflicts. In the mid-1980s, the balance of power in OPEC has shifted from the 'high absorbers' to the low absorbing countries, in particular the countries of the Gulf Cooperation Council. The fundamental difference of opinion between these two groups in OPEC makes policy-making increasingly difficult. OPEC has entered yet another phase in its existence. Its role as a cartel, at least in its present formation, has structurally changed. OPEC will have to seek a new position in the market, just as the major oil companies were forced to do before them.

Chapter Six

The Major Oil Companies in the 1970s and 1980s

1. Segmentation of the international oil industry

The fundamental change in the international oil supply conditions[1] reverberated all through the industry. The dominant position of the large international companies, at least in crude oil reserves, supply and trade, was successfully challenged by the state oil companies of the oil producing countries. The former ceased to be the major producers of oil, at least in the short and mid term. The major oil companies' interest in crude oil production in the Middle East dropped sharply after 1973, when the host governments nationalized or achieved majority paricipation in crude oil production (See Figure 6.1). After 1980, the companies' equity crude interests diminished further, and their remaining holdings in OPEC countries were in sharp contrast to ten years earlier.

The backward disintegration unbalanced the companies in a far more serious way than the implementation of the MOICP had in 1958. Although the imposure of the US import-quota had impeded the companies' global market control and disrupted their global upstream/downstream balance, they were able to restore their balance in both hemispheres fairly quickly. They increased their exploration efforts and took over companies in the US. The crude-rich companies expanded their processing capacity in the growing non-US consumer markets. In 1970, the large international oil companies still controlled 53.3%[2] of crude oil production, 56.7% of crude oil processing and 61.8% of sales of oil products.[3]

In 1973, the balance between oil production and processing was again disturbed. Unlike in 1958, when the major oil companies were cut off from an important market, they were suddenly confronted with backward disintegration. Between 1974 and 1979, the remainder of their equity crude, and crude obtained from producer governments under 'buy-back' agreements, did soften the loss of crude ownership for the companies. At least, the

[1] From 1972 onwards the OPEC member states negotiated with the international oil companies to nationalise the concessions in their countries or to participate in the consortia. Iraq and Iran, for instance, nationalised the IPC and the Iranian consortium, while Saudi Arabia chose to steadily increase its participation in ARAMCO. The decision-making on the production level was by the end of the 1970s a matter of government policy of the OPEC countries.

[2] Calculated from *OPEC Annual Statistical Bulletin 1973*.

[3] G. Luciani, op.cit., p.11.

companies continued to have a guaranteed supply. In 1979, approximately 47% of the world crude supply (compared to 69.5% in 1973) was still being passed through the channels of the oil companies and their affiliates.[4] However, the Third Party market had already contracted from 22.5% of world supply in 1973 to 11.2% in 1979.

Before 1973, the large international oil companies had traded the oil produced in excess of their processing and marketing capacity to other oil companies and refiners. During the 1970s, the volume of crude that was sold through the channels of the international oil companies decreased, and crude was progressively marketed through long-term supply-contracts by state oil companies, through government-to-government contracts, and the spot-market. By 1979, 42.2% of the world oil supply was marketed through the producing countries' channels.[5]

The changes at the supply side of the market also allowed new traders to enter the industry, and most OPEC countries diversified their sales outlets.[6] In the 1980s, a typical OPEC country had 20 to 40 customers, including the ex-concessionaires, US Independents, European and Japanese (national) oil companies, oil companies from developing countries, refiners, traders, and governments.

After 1979, the volume of oil traded on the international (spot) markets increased, and approximated 50% of total traded oil in 1983.[7] The change in the crude trade configuration again illustrated the erosion of the market power of the large international oil companies.[8] In 1950, they had passed 92.8% of all traded oil through their own channels, and those of their subsidiaries and affiliates. By 1970 this had decreased to around 70%, and in 1980 had declined to around 47%. The share of the state oil companies of producing countries increased from zero in 1960 to 42% in 1980. Since total trade contracted from 34,185 thousand b/d in 1973 to 27,377 thousand b/d in 1988[9] (See Table 3.1), the international oil companies experienced a declining share in a shrinking market. This resulted in a large loss of oil volume traded for the companies. The break-up of the Third Party market during the oil price rise in 1979/1980[10], and the diminishing volume of equity crude, had concluded a long period of domination of oil production and trade of the major international oil companies.

[4] G. Luciani, op.cit., p.13.

[5] Ibid., p.13.

[6] R. Mabro, The changing nature of the oil market and OPEC policies, in: *OPEC Review*, vol. VI, no. 4, Winter 1982, p.325.

[7] T.M.A. Farrell, The world oil market 1973-1983, and the future of world oil prices, in: *OPEC Review*, vol. IX, no. 4, Winter 1985, p.407.

[8] Ibid., p.406.

[9] World oil trade started to decline from a peak level in 1979 (35,350 thousand b/d) until 1984 (24,120 thousand b/d), increased to 26,265 thousand b/d in 1986, declined again to 24,590 thousand b/d in 1987 and increased in 1988 to 27,377 thousand b/d. Various *BPSR* issues.

[10] See Chapter 4.

The companies were forced to embark on backward re-integration instead of their former forward integration. Moreover, to re-establish the balance between upstream and downstream operations, the backward re-integration had to be paralleled by restructuring the downstream industry. The price increases in 1973/1974 and in 1979/1980 had concluded a period of expanding demand for oil and oil products, and also caused a major shift in demand to the lighter end of the refined barrel. The oil processing industry was thus forced to both rationalize processing capacity and to adjust the remaining capacity to the new chemistry of demand.[11]

The break-up of the old concessionary system and the increasing involvement of the producing governments in direct supply to refiners and consumers urged the companies to engage in exploration outside OPEC, and to defend their markets. The efforts of the companies to explore for oil were mainly concentrated in the industrialized countries. Especially, the development of the Alaska oil fields and the North Sea was accelerated. However, the major oil companies had to compete for these concessions with the Independents and state oil companies.

At the end of the 1970s and early 1980s, the North Sea emerged as an important oil producing region. Between 1978 and 1988, oil production in Norway increased from 350 thousand b/d to 1,150 thousand b/d, and in the United Kingdom from 1,095 thousand b/d in 1978 to 2,665 thousand b/d in 1986. However, in 1990 UK production had decreased to 1,895 thousand b/d.[12] In 1984, the UK and Norway together were the fourth largest oil producers (after the Soviet Union, the US and Saudi Arabia), and the third largest net-exporter (after the Soviet Union and Saudi Arabia). Their position was partly due to OPEC's production policy.[13] However, compared to the largest oil producing and exporting countries, North Sea proven reserves are fairly small. In 1988, the proven reserves of the United Kingdom and Norway were respectively 4.3 thousand million barrels, and 10.4 thousand million barrels, only 0.5% and 1.1% of world proven oil reserves. In 1988, the r/p ratio for was 24.6 years for Norway, and only 5.2 years for the UK. Norway's ratio had in 1990 declined to 12.2, while the UK's ratio increased to 5.6.[14]

Still, the impact of the North Sea on the world oil market has been substantial. The oil companies had replaced part of concessionary OPEC oil with North Sea oil production. The North Sea offered many advantages to the companies. The investment climate in the United Kingdom and Norway was favourable, with the advantage of large nearby consumer market. The disadvantage of higher exploration and operating costs in the North Sea was compensated

[11] See J. Boers, G. Eisenloefel and C. van der Linde, op.cit.; W. Molle, op.cit., in: H.W. de Jong (ed.), *The Structure of European Industry*, Kluwer Academic Publishers, Dordrecht, second revised edition, 1988; and S. Sinclair, op.cit., 1984.

[12] *BPSR 1989.*

[13] R. Mabro, R. Bacon, M. Chadwick, M. Halliwell and D. Long, *The market for North Sea Crude Oil*, Oxford Institute for Energy Studies/Oxford University Press, 1986, p.58.

[14] *BPSR 1991.*

by tax incentives from the United Kingdom and Norway[15]. Investments were also assisted by OPEC policy to stabilise the oil price at a high level.

The licensees in the North Sea are a mixture of private and national companies. The licenses are distributed among major oil companies, US Independents, European oil companies, (former) national oil companies, and non-oil companies (banks, insurance companies, holding companies and industrial corporations).[16] The number of North Sea operators is, however, much smaller, and is dominated by Shell, BP, Mobil and Phillips Petroleum. Their respective shares in the volume of operation were in 1984: 23.3%, 16.8%, 14.6% and 10.4%.[17]

In 1984, North Sea oil production was contributing 18.7% to BP's worldwide crude production, 9.4% of Shell's worldwide production, 24.9% of Exxon's worldwide production, 17.8% of Mobil's worldwide production, 3.1% of Chevron's worldwide production, and only 0.1% of Texaco's worldwide production.[18]

Due to the oil price increases, the prospect of producing oil from former marginal fields had improved substantially. Especially after the oil price increases of 1979/1980, world oil exploration soared. The number of rigs employed increased enormously, from 4,942 active rigs in the world (excluding CPEs) in 1980 to 6,221 rigs in 1981, an increase of 25.9%.[19] In the US the number of rigs employed increased 32.7% in 1981/1980 (3,613 to 4,797 rigs), but dropped 35% in 1982 (to 3,120 rigs). Since 1982, the (nominal and real) price decrease has slowed exploration and the number of rigs employed in the world has dropped substantially. In 1987, only 2,346 rigs were employed in the world (excluding CPEs).[20] OPEC employed 186 active rigs, 7.9% of the number of active rigs in the world (excluding CPEs). North America employed 56.8% (1,333) of the active rigs, and Latin America 15.1%, Western Europe 7.8%, the Middle East 4.2%, Africa 4.7%, Oceania 1.0%, and Asia and the Far East 10.3%.

[15] "The oil fiscal regimes in both the UK and Norway allow for the cumulative deduction of capital costs (within an uplift) from revenues before the payment of certain taxes. This system ensures that the margin is much greater on oil from non-tax paying fields than from any other source, and hence that crude from these fields will be fully utilized before other sources of supply are brought into operation. (This will be subject of course to the usual constraints imposed by good field management and by government regulations on best-practice techniques, gas flaring and other relevant factors.) Indeed the North Sea tax systems, compared with the familiar alternative which provides for a depreciation allowance over the notional lifetime of the assets, tend to shorten the period over which capital expenditures are recouped. For fields which have become fully taxpaying the margin is clearly smaller since there are no capital-related tax offsets to claim. In the present conditions of the oil market the North Sea costs structure and fiscal regimes provide firms with relatively high margins compared to other sources, so the firms will tend to take as much from these fields as possible." R. Mabro et al, op.cit., 1986, p.72-73.

[16] Ibid., p.30.

[17] Ibid., p.31.

[18] Ibid., p.33.

[19] *OPEC Annual Statistical Bulletin 1984*, Table 36.

[20] *OPEC Annual Statistical Bulletin 1987*, Table 35.

On top of the 'protective umbrella' of OPEC pricing policy, exploration was also stimulated by the governments of the oil-importing countries. The oil price increase of 1979/1980, combined with the sharp increase in the dollar exchange rate, exacerbated many countries' balance-of-payments situations. Governments of the oil-importing countries wanted to lessen the dependence on imports in general, and from OPEC countries in particular.[21] Both on a national level and on an international level, OECD governments pursued energy diversification and conservation policies.[22] Between 1979 and 1988, oil imports in the US and Western Europe declined drastically (See Figures 3.7 and 3.8). After 1980, world exploration expanded, and the number of oil producing countries increased to about 75.[23] However, most of the new producing countries remained small producers, but they at least managed to lessen the dependence on oil imports.

2. Exploration and production

The large international oil companies competed for new NOPEC-concessions to restore their crude production level. Some of the former consortia companies had turned from crude rich to crude short companies. From 1983 onwards, Exxon, Mobil, Chevron and Texaco produced less oil than was processed (See Table 6.1). Until 1980, these companies had access to (diminishing) equity crude, and had 'buy-back' agreements with the Saudi government. After 1983, the former ARAMCO owners maintained preferential access to ARAMCO oil.[24]

The collapse of long-term trading in 1979 affected the official crude positions of the majors significantly. The disruption of supply from Iran during the Islamic Revolution in 1979 forced BP, Exxon and Socal (Chevron) to declare 'force majeure' on a large number of their long-term contract customers.[25] BP lost close to 1.4 million b/d, Exxon 400,000 b/d and Chevron 300,000 b/d. A number of refiners that had obtained the crude through long-term contracts from the oil companies went to the spot market or dealt directly with the

[21] In 1973, the share of oil in primary energy consumption in the European Community was 62.1% of which 97.8% had to be imported. The geographic concentration of these oil imports was very large. The oil imports of France originated mainly from Algeria (20%) and Libya (17%), while 41% of the oil imports of Germany were from Libya, 53% of the oil imports in Greece were from Iraq, 30% of Italy's imports were from Libya and Ireland depended on imports from Kuwait (30%) and Saudi Arabia (27%). R. Lefeber and J.G. van der Linde, op.cit., p.455.

[22] Most of these countries joined the International Energy Agency. France is a noteworthy exception. J.G. van der Linde and R. Lefeber, op.cit., p.12.

[23] T.M.A. Farrell, op.cit., p.406.

[24] Until 1973, the geographic diversification of the ARAMCO concessionairies (except Exxon) was fairly limited. Production capacity of their Middle Eastern oil assets already created a surplus over their processing capacity. Actively competing for new concessions would only have created a larger surplus. Moreover, in the 1960s, the new producing countries, like Libya, offered smaller concession areas, which could be obtained on less attractive terms.

[25] Y. Stournaras, op.cit., p.44-45.

oil producers, since they were willing to pay a premium to obtain crude oil. As a result, the system of long-term supply contracts, which passed the crude through the channels of the major oil companies, broke up and was replaced by short-term contracts or spot market trading.

The major international oil companies did, however, not lose all their concessionary OPEC oil. Some OPEC countries preferred majority participation, and in some cases minority participation only, but in general the stake of the international companies in oil production in OPEC countries was minimized. In 1988, Exxon still had producing rights in Indonesia (50% of P.T. Stanvac) and the United Arab Emirates (4.75% in ADCO).[26] Mobil retained producing and production sharing rights in Indonesia (50% in P.T. Stanvac and 100% production sharing), in Nigeria (40% in Mobil/NNPC and 50% in Tenneco) and in the United Arab Emirates (4.75% in ADCO). Chevron remained a 50% owner in the Indonesian companies P.T. Caltex and Calasiatic/Topco (joint-ventures with Texaco). Chevron also had a 40% interest in the Nigerian company Gulf/NNPC, while Texaco had a 60% interest in Texaco/NNPC. BP only had interests in ADCO (24%) and ADMA (37%) of the United Arab Emirates. In 1988, Shell still had a 75% interest in Shell-Gabon of Gabon, a 20% share in Shell/NNPC of Nigeria and a 9.5% share in ADCO of the United Arab Emirates.

The Independents and some state oil companies from consuming countries also managed to preserve their interests in some OPEC countries. In Algeria, Medaloil, Hispanoil, and Braspetro kept 49% minorities shares in new joint-ventures with the government or the national oil company. CFP had a 85% share in Total Algerie, a 100% production sharing agreement in Indonesia, and a 4.75% share in ADCO of the United Arab Emirates. In Gabon, ELF and AMOCO had a 75% share in a producing enterprise. In Kuwait, the Arabian Oil Company (Japan) had a 80% interest in a joint-venture with the government. In Libya, OASIS, Occidental, OMV, Agip, Aquitane, Gelsenberg, and Gracepetroleum all had minority shares in joint-ventures with the government or the national oil company. Wintershall is still producing as an independent producer in Libya. In Nigeria, Agip, Phillips, ELF and Tenneco had minority shares, while Phillips and Tenneco also had full subsidiaries in Nigeria. Ashland had a production sharing agreement with Nigeria. In Saudi Arabia, the Arab Oil Company had a 80% in a joint-venture with the government, and Getty Oil is still a full subsidiary (now of Texaco) with producing rights. In the United Arab Emirates, TOTAL, Amerada Hess, Buttes and Zadco (a 50% subsidiary of CFP) had producing interests.

The international oil companies acquired new oil production by means of two strategies. They invested in new oil producing areas in the US and Europe (notably Alaska and the North Sea), but also in Asia (China, Malaysia, Brunei), in the Middle East (Oman, Yemen), in Africa (Sudan, Egypt, Angola) and Latin America (Peru, Colombia). In some instances the experiences of the companies were not so positive. Oil production in Peru, Colombia,

[26] *OPEC Statistical Bulletin 1988*, Table 43.

Angola and Sudan is troubled because of internal strife in the countries. Texaco had to abandon its operations in the south of Sudan because of the civil war in the country in the mid-1980s. Shell was more fortunate in Brunei and Malaysia, but is having trouble with its operations in Colombia and Peru. However, Shell remained a company with a very diversified production strategy.[27] BP withdrew almost from global oil exploration and production, depending almost entirely on oil production from Prudhoe Bay in Alaska and its fields in the North Sea.

Other ways of acquiring oil reserves are through take-overs of existing companies and through share buy-backs.[28] The 1980s have been very dynamic in terms of mergers and take-overs. The cost of finding oil through exploration has been higher than the cost of a barrel of oil through take-overs. Although the average cost of finding new oil has declined since 1985 from $10 a barrel to $5 a barrel, it was still higher than oil on 'Wall Street', which averaged $4.50 a barrel in 1988.[29] Chevron took over Gulf[30] (except its downstream operations in some European countries which were sold to Kuwait Petroleum Company), Texaco took over Getty Oil[31], Shell purchased, among other acquisitions[32], the remaining 30% of the shares in Shell Oil (listed as the fifth largest American oil company) in 1985, and BP acquired Sohio, a former minority holding of the company and Britoil. Exxon spent, besides the $7 billion on buy-back shares, $1.4 billion on oil properties in the period 1983-1987. In 1988 alone, a total amount of $26.1 billion was spent on oil and gas asset purchases; 32.6% of these purchases were in Western Europe, 30.3% in Canada, 28.3% in

[27] In an article in *The Wall Steet Journal* (July 9, 1987) the strategies of Exxon and Shell are compared: "The assumptions underlying both companies' strategies are similar. Both are betting on an eventual resurgence in the oil business. And both have cut employment, refining capacity and capital spending since oil prices began sliding a few years ago. But Royal Dutch/Shell is pushing for market share while Exxon has pulled back. (....) While Exxon cuts oil exploration, chiefly in the U.S., Royal Dutch/Shell presses its search for new fields, Mr. Holmes says. As one result, Shell Oil Co., the group's U.S. unit, passed Exxon last year as the leading oil-well driller in the U.S. Royal Dutch/Shell has also moved to fill the void left by Exxon's pullback from some markets."

[28] Since 1983, Exxon spent more than $7 billion buying back its own shares. According to Exxon, it helps to raise the per-share earnings and the price of the stock. The acquired oil reserves (their own) are thus purchased at a price of $2.50-$3.00 a barrel. *The Wall Street Journal*, July 9, 1987.

[29] *The Economist*, May 6, 1989, p.75.

[30] The price of Gulf's oil reserves at barely $4 a barrel versus an average cost, in 1983, of more than $10 a barrel for finding new U.S. oil reserves, made the company at a market capitalization of only 2.2 times cash flow very vulnerable to a take-over. The actions by T.Boone Pickens of Mesa Petroleum Company forced Gulf to merge with Chevron.

[31] The litigation between Pennzoil and Texaco resulted in an out-of-court settlement of $3.0 billion in cash. Texaco purchased Getty Oil for $128 a share. See for a detailed description of this take-over: T. Petzinger, *Oil & Honor The Texaco-Pennzoil Wars, inside the $11 billion Battle for Getty Oil*, G.P. Putnam's Sons, New York, 1987; See for an excellent analysis of the litigation and subsequent settlement: R.H. Mnookin and R.B. Wilson, Rational Bargaining and Market Efficiency: Understanding *Pennzoil v. Texaco*, in: *Virginia Law Review*, Vol. 75, 1989, p.295-331.

[32] Shell spent since 1983 $11 to $12 billion on acquistions, including the $5.5 billion acquisition of the 30% remaining shares in Shell Oil Co. *The Wall Street Journal*, July 9, 1987.

the US, and 8.8% in the rest of the world.[33]

In 1988-1989, the equivalent 2.3 billion barrels of North Sea oil and gas changed hands (almost a sixth of Britain's known reserves) through take-overs against an average price of $3.80 a barrel. The North Sea assets were attractive to large oil companies, because 84% of their oil exploration and appraisal costs could be set against the Petroleum Revenue Taxation. The assets of smaller North Sea companies[34] were sold because of dwindling cash flow and because most companies were not typical (integrated) oil companies and did not invest in oil exploration with the long view the integrated companies take. As a result the North Sea assets have become concentrated to a few companies. The Economist[35] has predicted that half of Britain's oil production will be concentrated in BP, Shell and Exxon in the next three years. By itself, BP owns a fifth of Britain's oil and gas reserves and production.

Lately, the possibilities for companies to acquire cheap oil reserves on Wall Street have diminished. Only a few companies had, under the present circumstances of depressed oil prices, the available cash to acquire new assets. Some companies were in a precarious credit situation because of low oil prices; others were still investing in upgrading their refinery capacity. Only Shell and Exxon were recorded to have a 'problem' investing their enormous cash-flows efficiently. If oil prices remain depressed for a longer period, and smaller (American) companies are not able to compensate profits and losses in either their upstream or downstream sector, increased concentration will be the result. *The Economist*[36] quotes industry specialists who expect increased concentration in the oil industry as a result of take-overs. Some argue that quite a lot of 'second tier' companies, like Unocal (no. 14 on the list of the *Oil and Gas Journal* 400), Texaco (no. 3), Phillips Petroleum (no. 11), Occidental Petroleum (no. 9), Amerada Hess (no. 20), Kerr-McGee (no. 28), and Texas Eastern (no.19) will be taken over by competitors, because they have a high net-asset-value relative to their current share prices. Other experts expect the number of big integrated oil companies to halve by the turn of the century.

Another problem facing the industry is that the acquisitions of 'Wall Street oil' reduced spending on exploration, resulting in a run down of the total available reserves. In 1988, only half of the large oil companies' oil and gas production had been replaced by new discoveries.[37] Restrained exploration spending has also affected the oil services industry, forcing companies to scrap aging material and cut capacity as a result. In time, when

[33] *The Economist*, 6 May 1989, p.76.

[34] BP acquired Britoil at $2.81 a barrel; ARCO (Atlantic Richfield) acquired Tricentol at $3.23 a barrel; Elf Aquitaine acquired RTZ Oil and Gas at $4.76 a barrel; British Gas acuired Acre Oil at $5.17 a barrel; Amerada Hess acquired Whitehall Petroleum at $3.24 a barrel; Ultramar acquired Blackfriars Oil and Gas at $3.90 a barrel; Lasmo acquired Thomson North Sea at $3.59 a barrel; and Enterprise Oil acquired the North Sea interests of Texas Eastern at $3.35 a barrel. *The Economist*, March 18, 1989.

[35] Ibid.

[36] Ibid., 6 May 1989, p.75.

[37] Ibid.

exploration spending does increase, the equipment costs could rise because of capacity restraints. This could increase the cost of finding new oil.

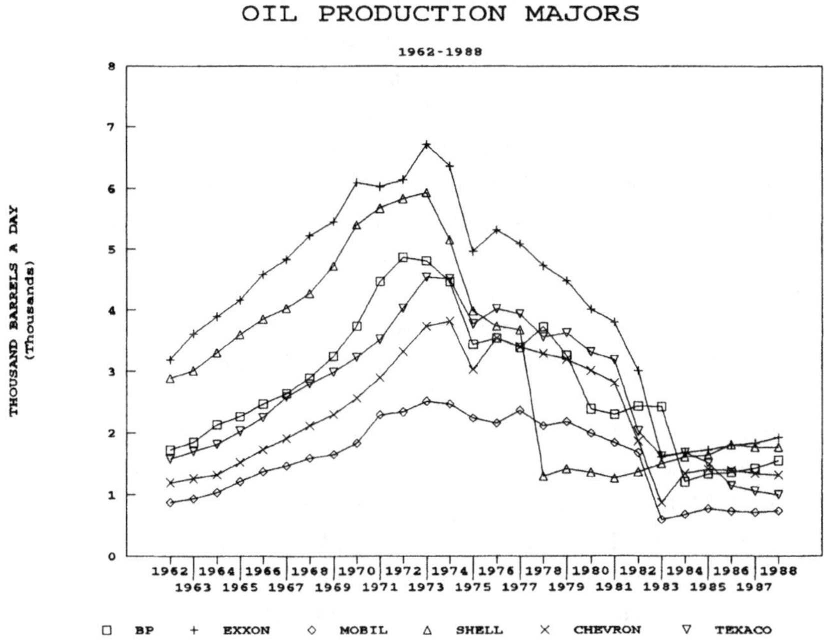

Figure 6.1

The group of six (until 1984 seven) companies' share in world oil production declined from 54.8% in 1973 to 30.1% in 1979, and 13.3% in 1988 (See Figure 6.1).[38] Until 1973, major company oil production increased considerably, and was comparable to the expansion of world oil production. In 1972, Iraq nationalized the IPC, and the shareholders lost this part of their equity crude. Between 1973 and 1975, crude production of the majors was affected by nationalizations, participations, and a decrease in production because of the general economic situation in the world. All companies, except Mobil, lost a substantial share of their production. In 1979/1980, the companies again lost a substantial share of their crude oil production. The main cause for the American companies' loss of oil production was the 100% participation of the Saudi government in ARAMCO. The loss of their long-term contracts with the state oil companies or governments was another important reason for the decline in production. The loss of Saudi crude to the former ARAMCO partners has changed

[38] Production of oil companies includes oil obtained under long-term contracts and oil produced by subsidiaries, affiliates and minority shareholdings. The American oil companies' production sometimes includes natural gas liquids. The disparities between the production figures for the companies do not alter the general trend.

their position distinctly. However, the companies maintained their strong links with ARAMCO, and they still handle a substantial share of Saudi crude. This crude is considered 'safe' crude, although these links were not secured by long-term contracts.[39] The lack of a large processing sector and an international distribution network of the oil producing countries together ensure the major oil companies (at least on the short-term) sufficient crude.

The major oil companies (See Table 6.1) have not replaced their lost production by new oil production. Their share in NCW oil production dropped from 51.4% in 1974[40], to 36.9% in 1980, and to 17.8% in 1988.[41] In 1987, Exxon was producing 45.8% of its 1980 production level, and 28.8% of its 1974 production level. The fall in production for Mobil, Texaco and Chevron is even larger.[42] Only Shell was producing more oil than in 1980, but still less than in 1974. Shell has always been a rather crude-short company and had, before 1973, developed more geographically diversified production activities.[43] In 1974, 41.1% of Shell's supply was from the Middle East, compared to 51.3% of Exxon's supply (includes Africa), 81.5% of BP's supply, 61.1% of Texaco's supply, 70.4% of Chevron's supply, and 59.9% of Mobil's supply. Between 1974 and 1980, the dependence on Middle Eastern supplies of the American oil companies hardly changed.[44] In 1980, Exxon still obtained 55.1% of its worldwide supplies from the Middle East and Africa, Mobil 58.6%, and Chevron 81.1%. In 1980, Shell and BP had already substantially diminished (out of free will or not) their dependence on Middle East supplies, to 25.2% and 18.4% respectively, and had already diversified their production to other oil producing regions.[45] In 1980, Gulf's dependence on supplies from the Middle East had decreased to 16%.

[39] "The exact nature of the agreements among the Saudi government, Aramco and the four 'partners' is not altogether clear. Although Aramco is owned by the Saudi government-- in other words, it is called on to implement the government's decision and policies-- at the same time, there is a continuing 'closeness' with the four partners.(...) In practice, Aramco normally sells its oil to the four partners in the same proportions as before (Mobil gets somewhat more than its 'equity' share)." Aramco gets the maximum production minus the oil that Petromin sells directly and the oil that is allocated to specific companies by the government (incentive crude to companies in downstream joint-ventures in Saudi Arabia, estimated at around 900,000 b/d). G. Luciani, op.cit., p.39-40 and 120.

[40] Calculations based on *OPEC Annual Statistical Bulletin*, various editions; *Company Annual Reports*, and *BPSR 1989*.

[41] Total oil production of the seven in 1974 was 30,092 thousand b/d. Total oil production of the six (seven until 1983) companies in 1980: 20,435 thousand b/d; in 1981 19,010 thousand b/d; in 1982 16,215 thousand b/d; in 1983 12.514 thousand b/d; in 1984 12,510 thousand b/d; in 1985 12,574 thousand b/d; in 1986 12,969 and 1987 13,449 thousand b/d.

[42] In 1987, Mobil was producing 35.6% of 1980 production and 28.8% of 1974 production. Texaco was producing 36.2% of 1980 production and 26.7% of 1974 production. Chevron was producing 44.5% of 1980 production and 35.1% of 1974 production.

[43] The number of countries in which Shell had upstream interests was 51 in 1984, while it produced oil and/or gas in 27 countries.

[44] G. Luciani, op.cit., p.43-42.

[45] The recuperation of BP production was strongly correlated to the development of North Sea and Alaska oil operations.

Table 6.1 Large oil company oil production and processing

	1980	1981	1982	1983	1984	1985	1986	1987	1988	1989
BP										
production	2386	2298	2433	2425	1198	1328	1351	1425	1550	1412
processing	1917	1638	1550	1604	1700	1795	1895	1822	1924	1244
balance	469	660	883	821	-502	-467	-544	-397	-374	168
Exxon										
production	4008	3796	3009	1607	1678	1720	1796	1835	1919	1804
processing	4149	3878	3496	3266	3220	2903	3032	2981	3042	3299
balance	-141	-82	-487	-1659	-1542	-1183	-1236	-1146	-1123	-1495
Shell										
production	1352	1260	1366	1497	1609	1639	1803	1766	1763	1851
processing	3382	2955	2971	2927	3048	2959	3003	3035	3105	3129
balance	-2030	-1695	1605	-1430	-1439	-1320	-1200	-1269	-1342	-1278
Mobil										
production	1991	1844	1674	591	675	772	727	709	735	761
processing	1959	1773	1711	1594	1620	1590	1545	1545	1563	1646
balance	32	71	-37	-1003	-945	-818	-818	-854	-911	-885
Chevron										
production	3009	2813	1863	864	1335	1399	1389	1338	1309	1308
processing	2186	1988	1544	1433	2036	1956	1800	1878	1907	1893
balance	823	825	319	-569	-701	-557	-411	-540	-598	-585
Texaco										
production	3316	3188	2025	1615	1678	1508	1135	1056	990	833
processing	2533	2211	1795	1708	2025	1848	1934	1892	1877	1356
balance	783	977	230	-93	-347	-340	-799	-836	-887	-523
Gulf										
production	1615	1393	1420	1499						
processing	1385	1133	1027	939						
balance	230	260	393	560						
total production	17677	16592	13790	10098	8173	8366	8201	8129	8266	7969
total processing	17511	15576	14094	13471	13649	13051	13209	13171	13501	12567

production: produced, net equity production only; total production: total production of six (seven) companies; total processing: total processing of six (seven) companies; source: OPEC Statistical Bulletin 1984, 1988 and 1989, Tables 72; Royal Dutch Shell Group, Financial and Operational Information 1985-1989, 1969-1978, 1978-1982 and 1983-1985, and own computations.

In 1988, Shell[46] produced 21.3% of the group's (the former Seven Sisters) oil, and Exxon 23.2%. BP's share was 18.8% of the groups' total production, and Mobil's 8.9%, Chevron's 15.8%, and Texaco's 11.8%. The companies' share of world oil production (13.3% in 1988)

[46] Shell already relied on long-term contracts in the 1960s, and continued to purchase its oil from Third Parties in the 1980s. In Shell's *Financial and Operational Information 1985-1989*, the company's net equity production is defined as: "Of Group companies, plus Group share of associated companies, and including natural gas liquids. Royalty purchases are excluded. In those countries where production sharing contracts operate, the figures shown represent the entitlements of the Group companies concerned under these contracts." Under the heading Purchases (defined as:" By Group companies from third parties plus/minus purchases from associated companies greater than/less than the Group share of net production.") supply was: 1985 2,788 thousand b/d; 1986 3,226 thousand b/d; 1987 3,724 thousand b/d; 1988 3,945 thousand b/d; and 1989 4,026 thousand b/d.

compares to the production level of some regions or important producers. In 1988, for instance, Africa produced 8.7% of total world oil production, Latin America 11.3%, the US 15.3%, and Saudi Arabia 8.5%.[47] Until 1973, the major oil companies were, with the exception of Shell, extremely crude rich, and had access to low cost oil for a prolonged period.

Between 1980 and 1988, none of the companies produced more oil than was being processed. From 1984, BP had only a small deficit of production over processing. In 1989, BP's self-sufficiency[48] had declined to 73%. The initial substantial deficit of Exxon and Shell diminished in the course of the 1980s. Shell's rate of self-sufficiency steadily increased from 39% in 1981 to 55% in 1989.[49] Exxon's rate was 52% in 1989. Chevron (despite the merger with Gulf[50]), Mobil and Texaco also experienced a deficit in processing over production in the 1980s.

The companies' renewed efforts to improve their crude production capacity showed in their capital expenditure. Capital expenditure on exploration and production (See Table 6.2) in (nominal) millions of dollars clearly distinguishes BP, Exxon and Shell from the other 3 majors. BP's, Exxon's, and Shell's capital expenditures on exploration and production was substantial, compared to the other companies' expenditures.[51]

In 1980, BP, Exxon, and Shell spent 15.2%, 14.3% and 14.3%, respectively, of their total capital expenditure on exploration.[52] That year, Mobil spent 10.9% of total capital expediture on exploration, Chevron 17.5%, and Texaco 38.8%. In 1988, BP spent 71.4% of total capital expenditure on exploration and production (compared to 61.7% in 1980), compared to Exxon and Shell 53.3% and 49.5%, respectively. Mobil, Chevron, and Texaco, spent 49%, 45.4%, and 61.2%, respectively.

The oil price collapse in 1986 negatively affected expenditure on exploration and production. BP's expenditure declined with 19% in 1986/1985, but increased 11.9% again

[47] *BPSR 1989.*

[48] Self-sufficiency is defined as as net equity oil output divided by refinery runs. *Petroleum Intelligence Weekly*, June 11, 1990, p.5.

[49] Ibid.

[50] Gulf sold the downstream operations in some countries in Western Europe to KPC. Chevron did, however, take over downstream operations of Texaco in the Netherlands. Chevron has always had a crude rich position in the eastern hemisphere market.

[51] Shell's capital expenditure on exploration and production was comparable to Exxon's in the 1960s and high compared to the other five companies. *OPEC Annual Statistical Bulletin 1966 and 1969.*

[52] *OPEC Annual Statistical Bulletin 1984 and 1988*, Table 75. Total capital expenditure in 1980 and 1988 in million dollars: BP 4,131 and 10,383; Exxon: 8,040 and 7,500; Gulf: 3,001, merged with Chevron in 1984; Mobil: 4,815 and 3,915; Shell: 7,959 and 8,160; Chevron: 3,599 and 2,405; Texaco: 2,691 and 2,611. Capital and exploratory expenditure peaked in: 1988 for BP; 1985 for Exxon (10,800); 1981 for Gulf (4,325); 1982 for Mobil (5,253); 1981 for Shell (9,320); 1981 for Chevron (5,568), and 1982 for Texaco (3,768).

Table 6.2 Company capital expenditure on exploration and production 1980-1989 in million US$

	BP	Exxon	Shell	Mobil	Chevron	Texaco	Gulf
1980	2,547	5,213	4,979	2,285	2,230	2,043	2,239
1981	3,952	7,346	5,911	3,009	3,562	2,184	3,457
1982	4,549	7,330	5,487	3,330	2,590	2,620	2,852
1983	3,350	6,085	4,600	2,556	2,090	2,367	2,340
1984	3,412	7,023	3,732	2,445	*3,346	2,532	-
1985	3,565	7,561	5,021	2,308	2,802	2,100	-
1986	2,889	4,600	2,833	3,540	1,704	1,382	-
1987	3,240	3,300	2,744	1,602	1,763	1,247	-
1988	7,417	4,000	4,040	1,919	1,092	1,599	-
1989	3,341	4,200	4,975	1,787	2,079	1,156	-

*merger between Chevron and Gulf; source: OPEC Annual Statistical Bulletin 1984, 1987 and 1989.

in 1987/1986. Exxon's expenditure decreased 39.2% in 1986/1985, 28.3% in 1987/1986, and increased 21.2% in 1988/1987. Shell's expenditure fell from 6,635 million dollars in 1985 to 4,314 million dollars in 1986 (35%), to 3,713 million dollars in 1987 (13.9%), and increased to 4,040 million dollars in 1988 (46.2%). Mobil, Chevron and Texaco experienced similar declines in expenditure. From 1985 to 1987 Mobil decreased expenditure by 30.6%, Chevron by 37.1% and Texaco by 40.8%.

Within a group of selected large American independent companies, only AMOCO can compare in total capital expenditure to the expenditure of the second tier of the large oil companies in 1986 and 1987 (See Table 6.3). In 1987/1986, the major companies, except for Shell and BP, decreased expenditure. The decrease of Texaco's expenditure can be partly explained by the problems the company was facing, pending the decision of the court in the Penzoil case. Net income of Texaco was negative in 1987 (-4,407 million dollars). The increase in total capital and exploration spending of the smaller Independents is striking. Considering the general decrease in exploration and production capital spending[53], the increase must be accounted for by spending in the processing phase of the industry. In the US, the process of restructuring the processing industry, especially by the smaller companies, has commenced much later than in Western Europe. Government protection of smaller refiners has delayed the upgrading of refinery capacity.[54]

[53] *Oil & Gas Data Book 1989*, p.5. Capital and exploratory spending decreased 5.46% for the 400 largest US companies. The ratio capital spending to net income was about 4.4 to 1 in fiscal year 1987 compared to 8.5 to 1 in 1986. The decrease of the ratio was, however, mainly the result of higher income rather than lower spending.

[54] See Chapter 1.

Table 6.3 Total capital and exploration expenditure in 1986 and 1987 for a selected group of companies in million US$

	1986	1987	% change
Exxon	7,200	7,136	-0.9
Shell*	6,857	7,218	+5.3
BP*	5,567	5,832	+4.8
Mobil	3,046	2,942	-3.4
Chevron	3,018	2,841	-5.9
Texaco	2,369	1,990	-16.0
Amoco	3,181	2,979	-6.4
Occidental	1,618	757	-53.2
Arco	1,822	1,463	-19.7
Tenneco	1,188	1,291	+8.7
Conoco	1,410	1,480	+4.9
Phillips	655	750	+14.5
Ashland	314	388	+23.7
Amerada Hess	366	454	+24.0

source: Oil & Gas Journal, September 21, 1987; The Oil & Gas Data Book 1989; * OPEC Annual Statistical Bulletin 1987, Table 75.

In the period 1977-1986, worldwide capital spending on exploration and production has shown, like the six companies' spending, a strong correlation with fluctuations in the crude oil prices.[55] In 1977, total expenditure was $34 billion, and it increased rapidly to $91 billion in 1981, decreased to $77 billion in 1985, and dropped to $54 billion in 1986. From 1984 to 1985, spending decreased by 7.8%, and in 1986/1985 26.1%. Except for BP, the major oil companies' expenditures declined a larger percentage than worldwide expenditure, but less than the percentage decline in US' expenditure (43.1% in 1986/1985).

Worldwide upstream spending did, however, blur the regional disparities in spending. In the US, spending has fluctuated from $17 billion in 1977, to $55 billion in 1985, and $25 billion in 1986. The strong correlation to the oil price development between 1977 and 1986 and upstream capital expenditure is explained by Chase as follows: "(...) in the United States, the private companies are the driving force for capital spending. The dynamics of these private companies are straightforward- investment is largely internally financed, so that when prices increase and cash flow improves, the funds are plowed back into capital projects. When prices decline, however, the process is reversed."[56]

The oil price decline, and the subsequent decline in capital expenditure in the American

[55] *Capital Investments of the World Petroleum Industry 1985 and 1986*, Global Energy Component, Chase Manhattan Bank, New York, 1988, p.5.

[56] Ibid.

oil industry, has affected the share of the oil industry's capital expenditure within total US capital expenditure. This share has always shown a strong correlation with (real) oil price developments. From 1961 through 1971, when oil prices decreased in real terms, capital expenditure in the oil and gas industry fell from approximately 10% to 4% of total capital expenditure in the US economy. In the period 1973 to 1981 this share increased to 18%, and after 1982 declined again to less than 7% in 1986.[57]

In Western Europe (mainly the North Sea), upstream capital investments appeared to be less affected by changing oil prices. Expenditure increased from $7 billion in 1977 to a plateau of $11-$12 billion from 1982 to 1986. According to Chase, the more conducive government policies, the revision of North Sea tax policies, cost cutting, innovation, and the long lead times have been helpful in retaining upstream expenditure during this period.[58]

3. **Refining**

The oil price increases in the 1970s, and the subsequent segmentation of the international oil industry, resulted in structurally changed market conditions for the downstream industry. The refinery feedstock became more expensive, while the 'demand mix' also changed. A large part of refining capacity in the world became less adequate to meet demand, and a process of restructuring commenced.

The refining industry processes crude oil into a variety of oil products. Its output is a typical case of joint-production. The composition of output depends on the crude input (heavy/light and sour/sweet crudes), and the complexity of the refinery. The 'cut of the barrel' is determined by these factors. Refinery output is to a certain degree inflexible, once a refinery is installed. The long time-lag between the investment decision and operation of a refinery and the high fixed costs relative to variable costs compel the refiners to assess future supply and demand developments accurately and put a premium on 'security of supply'. The complexity of refining is adequately described in a *Shell Briefing Service* publication: "The skill in refining is being able to determine the volume and selection of crude oil and other feedstock intake so as to supply a variable product demand, at changing relative values, in a profitable manner. This is influenced by a number of factors, in particular the complexity of the processing plant available and on whether supplies are directed primarily at the local market or are also international".[59] Vertical integration safeguarded the refiners' volume, cost, and chemistry of supply. Refining capacity that was in operation in 1973 or came on stream shortly afterwards, reflected the structure of the oil

[57] Ibid., p.12-13.

[58] Ibid., p.5-6. The long lead times in developing projects in the North Sea might cause the upstream reaction to oil prices be delayed for some years and will reflect itself in a few years to come. The lead time for construction projects is ussually five years or more.

[59] *Shell Briefing Service*, no. 5, 1987, p.6.

industry prior to 1973. OPEC participation agreements, nationalization of oil production, and subsequently OPEC's price leadership, altered the very foundation of the refining industry.[60]

The loss of a large part of supply by the international companies included the loss of control over the mix of crudes and the determination of price differentials between heavy and sour crudes, and light and sweet crudes.[61] OPEC intended to counter the rapid depletion of light/sweet crudes and to export crude oil in accordance with the chemistry of reserves. New additions to reserves were mainly additions of heavier crudes, bringing the share of heavy oil of total reserves to approximately 85%.[62] In 1972 the price differentials (spot prices) for Middle East heavy crudes (31^0 API) and African light crudes (37-40^0 API) averaged only $0.98 per barrel, but increased to $2.67 per barrel in 1974. In 1978 and 1979, the price differentials increased even more sharply, when the gap increased from $1.95 to $5.03 per barrel.[63] Price differentials were also more frequently used for price shaving.

The (market) refineries of the international oil companies were, until the 1970s, mainly geared to process light sweet crudes. They are easier to produce and refine and yield more middle and light products than heavier crudes. The composition of the refined barrel reflected the then current demand for products in the sub-markets for final oil products and semi-finished oil products (See Table 1.1).

In 1970, the 'cut of the barrel' varied in the main consumer markets. In the US demand for gasolines was dominant, and refineries were equipped to refine light sweet crudes into a maximum volume of gasolines. Demand for fuel oil, an important product in Japan and Western Europe, was less strong, since electricity was generated mainly by indigenous coal and gas production. In Western Europe and Japan, electricity was predominantly generated by fuel oil. Refinery output was equipped to handle demand for the various oil products. Increasing demand for products at the lighter end of the barrel in both markets was met by increasing refinery capacity (See Figure 1.1).

The oil price increases structurally changed the cost/revenue relation of the refining and petrochemical industry. The higher costs could not be passed on to consumers in all sub-markets. Competition of near substitutes, and unequal price-elasticities in the sub-markets, prohibited the companies from increasing prices of refined products in proportion to crude oil price increases. Especially fuel oil was sensitive to substitution by coal, nuclear energy and gas. The drive of the governments away from (OPEC) oil put the oil companies in an awkward situation to balance production off-refinery. Between 1974 and 1979, demand for

[60] See for the impact of the oil price increases on the refining and petrochemical industry in Europe: J. Boers, G. Eisenloeffel and C. van der Linde, op.cit.

[61] A slight alteration in price differentials can produce substantial costs in the downstream industry because of the inflexibility of refineries to adapt input. The margins on refining can be further strained when demand for oil products does not concur with 'the cut of the barrel' when the refinery input is changed.

[62] *OPEC Bulletin*, March 1982, p. 1.

[63] H. Tahmassebi, Crude oil and product differentials; historical perpective and outlook, in: *OPEC Review*, volume VIII, no. 3, Autumn 1984, p.245.

fuel oil dropped in the US, Japan, and Western Europe, continuing to decline until 1987 when fuel oil demand picked up slightly.[64]

The refineries had to adjust their output to the new demand mix. A rationalization and upgrading process of refinery capacity became inevitable after the oil price increase of 1979/1980. Refinery throughput and utilization rates declined.[65] Secondary conversion and new refining techniques were used to increase output at the lighter end of the barrel. In Western Europe, primary distillation capacity decreased from 19.2 million b/d in 1979 to 14.8 million b/d in 1986, while conversion capacity steadily rose from 2.5 million b/d to 4.0 million b/d between 1979 and 1986.[66]

The composition of NCW oil product demand changed in favour of lighter products. In 1978, the share of gasolines in world oil product demand was 26.7%, for middle distillates 30.5% and for fuel oil 28%. In 1988, the share of gasolines in NCW oil product demand was 28.9%, for middle distillates 35% and for fuel oil 18.7%.

The developments of oil product demand did differ, however, in the various consumer markets. In the US, the composition of oil product demand was traditionally dominated by gasoline, but became more pronounced after 1979. In 1978, 38.1% of total US oil product demand was for gasolines, and increased to 42.2% in 1988. Fuel oil demand decreased from 17.8% of total oil product demand in 1978 to 9.2% in 1988.[67] In Western Europe, the share of gasolines in total oil product demand was 20.5% in 1978, increasin to 24.8% in 1988. The share of fuel oil had been 31% in 1978, decreasing to 20.4% in 1988. In Japan a similar development took place. The share of gasolines in total Japanese product demand was 17.4% in 1978, increasing to 20.9% in 1988, while the share of fuel oil decreased from 47.4% in 1978 to 27.8% in 1988. Oil product demand in the LDCs continued to grow throughout the 1970s and 1980s, for the full range of products.

In 1970, the major international oil companies were operating approximately 50% of NCWs refining industry, with a similar share in product sales. The change in the market conditions in the 1970s affected them seriously. In 1979, they produced 35% of world oil production, while they operated close to 40% of the world refinery industry and had a similar share in product sales. Their balance between upstream and downstream operations was clearly deteriorating (See Table 6.1). The majors were among the first in the industry to restructure their downstream operations. They decreased their refinery capacity, and upgraded the remaining existing plants. Compared to the 1970s, processing output has declined. Parallel to restructuring their market refineries, Shell, Exxon and Mobil invested in source refineries in joint-venture with state companies in Saudi Arabia.[68] In 1988, the

[64] *BPSR 1989*, p. 12.

[65] See Chapter 1.

[66] *Shell Briefing Service*, no. 5, 1987, p.7.

[67] Calculations based on *BPSR 1989*, p.12.

[68] J. Boers, G. Eisenloefel and C. van der Linde, op.cit.

large international oil companies as a group had more or less restored the balance between upstream and downstream operations. They produced 12,211 thousand b/d and processed 13,501 thousand b/d. In terms of the uptream/downstream balance, their 19.6% share in crude oil production compares well with their 23.2% share in refinery throughput. However, the individual companies were not in balance.

4. Revenue and net income of the majors

The development of total gross revenue of the major oil companies (See Table 6.4) discloses the effect of crude oil price development in the 1980s on gross revenue of the companies. The oil price collapse is reflected in a decrease in gross revenue for all majors. BP's gross revenue decreased 25% in 1986/1985, Exxon's 17.6%, Mobil's 17.7%, Shell's 20.4%, Chevron's 40.1% and Texaco's 31.4%. The slight improvement in oil prices in 1987 and better refinery performance were translated into a recuperation of gross revenue. BP increased gross revenue from 1987/1986 by 13.2%, Exxon by 8.6%, Mobil by 13.7%, Shell by 20%, Chevron by 7.2% and Texaco by 8.4%.

Table 6.4 Gross revenue of major oil companies 1980-1989 in million US$

	BP+	Exxon*	Shell*	Mobil	Chevron	Texaco
1980	49,301	110,191	82,610	63,726	42,919	52,485
1981	53,880	114,923	88,454	68,587	46,609	59,297
1982	52,584	103,559	85,901	64,108	35,943	48,019
1983	49,219	94,734	84,345	58,998	29,182	41,147
1984	51,943	97,288	98,492	60,474	47,367	47,935
1985	54,400	92,869	94,569	60,609	45,325	47,514
1986	39,371	76,555	81,404	44,795	27,331	32,679
1987	47,616	83,335	98,026	51,387	29,120	35,344
1988	46,884	88,563	99,445	54,361	28,857	35,138
1989	50,493	96,285	106,090	56,188	32,785	35,656

+: includes from 1984 onward earnings from equity interest and stockholding loss or gain; *: includes from 1984 onward earnings from equity interest; BP including Sohio (BP America); Chevron including Gulf from 1984 onward; Texaco including Getty Oil from 1984 onward; source: OPEC Annual Statistical Bulletin 1984, 1988 and 1989, Table 73.

In terms of total revenue (See Table 6.5) and net income in 1986, the divergences among the six majors and a selected group of (largest) American Independents are eminent. In terms of revenue, Exxon and Shell are by far the first ranking companies, outstripping the revenue of the Independents many times. In 1986, Shell's net income was 69.3% of Exxon's net income, and in 1987 Shell had almost equall income (97.9%). In 1986, Mobil's net income was only 26.3% of Exxon's net income, and 26% in 1987. In 1986 BP's net income was a mere 13.7% of Exxon's but increased to 47.1% in 1987. Comparingly, in 1986, Occidental's net income totalled only 3.4% of Exxon's net income.

Table 6.5 Total revenue and net revenue of group of selected companies in 1987 in million US$

	revenue	net income	% net income/ revenue
Exxon	83,335	4,840	5.8
Shell	81,442	4,740	5.8
Mobil	56,716	1,258	2.2
BP	45,228	2,281	5.0
Texaco	35,315	-4,407	-12.5
Chevron	29,097	1,007	3.5
Amoco	22,388	1,360	6.1
Occidental	17,746	240	1.4
Arco	17,608	1,244	7.1
Tenneco	14,957	-218	-1.5
Conoco	11,097	277	2.5
Phillips Petroleum	10,917	35	0.3
Ashland	7,311	133	1.8
Amerada Hess	4,784	229	4.8

source: *The Oil & Gas Data Book* 1989 and *OPEC Annual Statistical Bulletin*. Disparity in revenues majors between table 6.4 and 6.5 is due to adjustments of revenues for the period 1984-1988 in the *OPEC Annual Statistical Bulletin*. For the sake of the comparison between majors and independents, the unadjusted revenues of *OPEC Annual Statistical Bulletin* and *The Oil & Gas Journal* are presented.

Net income as a percentage of total revenue also compares positively for both Exxon and Shell. BP's net income as a share of revenue recovered substantially from the rather low shares in the 1980s.[69] Mobil and Chevron suffered rather low net returns compared to revenue. Texaco experienced a staggering loss. Of the Independents, AMOCO and ARCO (Atlantic Richfield) and, to a lesser extent Amerada Hess, performed well.

The majors' net income as a percentage of total revenue (1980-1987) (See Table 6.6) discloses the overall performance of the companies. Shell and Exxon, performed much better during this period than the other four companies, although BP seemed to have made a fair

[69] *OPEC Annual Statistical Bulletin 1987*, p.138.

recovery. The backward re-integration, and the qualitative forward integration of the companies, has made them less sensitive to oil price decreases. However, US oil company taxation has, after the oil price collapse, been a disadvantage for companies which had their main producing activities in the US.[70] The share of US production in Exxon's worldwide production was 43.3% in 1987, compared to 49.3% for Mobil, 56.2% for Chevron, 55.5% for Texaco. However, both BP and Shell have substantial interests in the US through their subsidiaries Sohio (BP America) and Shell Oil Co. In terms of oil production in the US, Sohio ranks first in *The Oil & Gas Journal* list of largest American oil companies, and Shell Oil ranks fourth; in terms of US reserves they rank first and third, and Shell Oil drilled by far the most net wells.[71] The (larger) US Independents produced--except for Amoco (48.8%), Phillips Petroleum (56%) and Occidental (23.6%)--predominantly in the US.[72]

In 1988 total revenue of the majors was: Exxon 88,563 million dollars; Shell 99,445 million dollars; Mobil 54,361 million dollars; BP 46,884 million dollars; Texaco 35,138 million dollars; and Chevron 28,857 million dollars. In terms of net income, Exxon exceeded Shell in 1988. Exxon's net income was 5,260 million dollars, and Shell's net income was 5,234 million dollars. The other companies' net incomes were: BP 2,154 million dollars; Mobil 2,031 million dollars; Chevron 1,768 million dollars; and Texaco 1,304 million dollar (after two successive years of heavy losses; 1,425 million dollars in 1986, and 4,407 million dollars in 1987).[73]

The income of the American oil companies, but to a lesser extent also BP and Shell through their subsidiaries (Sohio and Shell Oil), was more adversely affected because the margins on production in the US were heavily affected by the lower crude oil prices. The US government has, in general, taxed oil production at lower marginal rates, and the 1986 price collapse was better cushioned by the other governments than in the US.[74]

Some Independents were severely hit by the price decline in 1986, and experienced precarious credit positions, inducing them to deplete reserves to repay debts, at the expense of exploration.[75] The lack of compensation from profits in the downstream sector of the

[70] "In the 1990s oil companies will have to be even more choosy about how and where they spend their cash on exploration. America, as an oil patch, will continue to fall from grace, principally because America's oil company tax regime is highly unfavourable at oil prices below $20 a barrel." *The Economist*, May 6, 1989, p. 76.

[71] *Oil & Gas Data Book 1989*, p.9.

[72] Ibid., p.8-9.

[73] *OPEC Annual Statistical Bulletin 1988*, Table 73.

[74] *Capital Investments of the World Petroleum Industry 1985 and 1986*, Chase Manhattan, 1988, p. 9-11. "As indicated, foreign governments, with their higher marginal tax rates, absorbed nearly $7 of the price decline, while the U.S. government only absorbed $5. This tax cushion resulted in a more severe decline in cash margins in the United States than overseas. In 1986, in fact, the U.S. and foreign cash margins were nearly equal; because capital development costs per barrel are generally higher in the U.S. than overseas, this implies a serious deterioration in the rate of return for U.S. capital expenditures. Overall, this shift in economics means that oil development projects overseas will begin to look relatively more attractive."

[75] Ibid., p.12.

industry exposed these companies more to the oil price decreases.

Table 6.6 The majors' net income as a percentage of total revenue 1980-1989

	BP	Shell	Exxon	Chevron	Mobil	Texaco
1980	6.8	6.7	4.9	5.6	4.4	4.3
1981	4.0	4.6	4.2	5.1	3.5	3.9
1982	2.4	4.0	4.0	3.8	1.9	2.7
1983	3.2	4.9	5.3	5.4	2.5	3.0
1984	2.8	5.5	5.7	3.2	2.1	0.6
1985	1.6	4.1	5.2	3.4	1.7	2.6
1986	1.9	5.5	7.0	-5.2	-1.0	-4.4
1987	4.8	4.7	5.8	4.3	2.4	-11.7
1988	4.6	5.3	5.9	6.1	3.8	3.7
1989	6.9	6.2	3.7	0.8	3.2	6.8

source: OPEC Statistical Bulletin 1984, 1988 and 1989, Table 74.

During the 1970s and the beginning of the 1980s, many oil companies have invested the bulk of their cash into upstream operations, and not in downstream operations. The depressed oil prices since 1986 (and the slight recovery in 1989), has distressed the oil companies upstream profits, and the majority of the companies could not compensate these decreases with better performances in the downstream industry.

The effect oil prices fluctuations on income also deviates between the six majors, because of the diversity in the compensation of profits in refining, marketing and petrochemicals. A $1 drop in oil prices in 1988 would, according to Steven Butler of the Financial Times, result in: a 5.7% drop in income for Shell, a 7.5% drop for Exxon, a 12% drop for BP and Texaco would loose nearly 25% of its income.[76]

5. Conclusion

In the 1960s, the increase in supply from new producers and competition for markets had gradually impaired the majors' control of the international oil market. Production increased, and market prices decreased. In the process, OPEC countries saw their economic rents and profits reduced. In the 1960s, the companies' share in world oil production was about 50-54%, while their share in NCW production fluctuated around 63-65%.[77] Compared to the

[76] *Financial Times*, October 10, 1988.

[77] In the 1950s, the companies' share was about 85% of oil production, excluding the United States and the centrally planned economies. See Chapter 4.

1950s, increased competition from the Independents and expanding Soviet oil exports had diminished the companies' share in oil production.

The rapid expansion of demand and the declining oil prices in the 1960s had stimulated investment in refinery capacity instead of exploration expenditure. Thus, the major companies' market position was first challenged by new entrants in oil production, and then also in the refinery sector. In the second half of the 1960s, oil exploration was in general sluggish. Exploration was not attractive as long as prices approached operating costs of new wells. The market was in over-supply. However, low oil prices and expanding demand did stimulate downstream investments.

In the early 1970s, the market began to tighten. Demand had continued to increase, while there had hardly been new additions to capacity, and production was reaching capacity. The downward price trend of the 1960s changed into an upward trend. However, this trend was not translated into a revision of the posted prices or at least not sufficiently to satisfy the demand for price increases by the host governments.

In 1970, Khadaffi of Libya forced the companies' to increase prices, and a round of negotiations with other governments followed. In 1970 and 1971, price and participation/nationalization negotiations were conducted between the companies and OPEC. The negotiations with OPEC in the early 1970s were strained.[78] A combination of political circumstances and market conditions facilitated OPEC's appropriation of market power. OPEC was in 1973/1974 producing 64.8% of NCW production and 53.4% of world production.

The price increases in the 1970s were an incentive for increased oil exploration. Formerly marginal fields were brought into production in the 1970s, and in the 1980s these fields seriously eroded OPEC's market share. The price increase of 1973/1974, but more so the increase of 1979/1980, set a new cycle of over-production and price decreases, and the recuperation of refinery margins was put into motion. The high oil prices attracted new entrants into the market. Stabilization of oil prices at a high level after 1980 created a protective umbrella for companies to extract oil from marginal sources. OPEC's share in NCW production declined from 57.2% in 1980 to 40.5% in 1985, and its share in world production declined from 43.7% to 30% in 1985. OPEC eventually lost its power over the market.

The companies had lost part of their equity production after 1973/1974, when the OPEC countries nationalized their oil resources and/or increased their participation in the companies' operations in their countries. Thoughout the 1970s, the companies still handled the lion's share of OPEC crude through their processing and marketing channels, but this came to an end during the 1979/1980 crisis.

The oil price increase had also seriously affected the companies' margins on refining, necessitating production capacity rationalization. In addition, the price increase had triggered

[78] See A. Hamilton, op.cit., Chapter 4.

a change in the chemistry of demand, and refining capacity had to be upgraded to meet demand at the lighter end of the barrel.

In 1988, the oil companies were producing 13.3% of world production and 17.8% of NCW production. The market conditions in the 1970s and 1980s have affected the companies' structure. When the companies' upstream and downstream operations became separated, they tried to restore their self-sufficiency through exploration and acquisitions.

Mobil, Chevron and Texaco had traditionally been less geographically diversified than BP, Shell and Exxon. The nationalization of their assets in the Middle East resulted in a geographic concentration of their upstream operations. More than half of Mobil's, Chevron's and Texaco's oil production is now concentrated in the US. Also BP's oil production is largely concentrated in the Alaskan and the North Sea regions. Only Shell's and Exxon's oil production is, at present, geographically diversified and represents their continued global approach of the oil markets. The difference between Exxon and Shell and the second and third tier companies has become more pronounced in the 1980s. In terms of capital expenditure, and upstream and downstream operations, they stand out. These two companies are major participants in all markets and have not retreated to any particular regional market. BP holds the middle ground between these companies and the second tier companies.

Compared to the 1950s and 1960s, the companies have lost a major share of their market power. However, the statistical evidence is deceptive. The major companies possess the capital and technical skills, which OPEC still lacks, to increase production capacity, downstream capacity, and to market the products on the consumer markets. These assets may help the companies to secure future supply. Their technical skills and capital could make a return to upstream operations in OPEC countries feasible, although the terms will be different from the past. The incomplete vertical integration of both the major oil companies (rate of self-sufficiency in crude production compared to processing) and the major oil producing countries (lack of processing capacity) could also bring about joint-ventures in the downstream industry. An alignment between the major oil companies and oil producing countries would yet again restructure the industry. In the future, the majors might be the natural allies of the state oil companies. Together, the majors and Saudi Arabia, Kuwait, Venezuela, and the UAE operated 30.3% of world oil production and 40.7% of NCW production in 1988.[79] Moreover, the countries are important exporters, and the majors have the processing capacity and the distribution network to channel the oil to the markets as long as the countries lack these facilities themselves.

In the 1980s, OPEC policy has been at odds with the interests of the crude-rich oil and/or integrated producing countries. At present, these countries (Saudi Arabia, Kuwait, UAE,

[79] Some double accounting could influence these percentages. However, in 1988 OPEC production was subject to quotas, and the countries were not producing at full capacity. Between 1988 and 1990, these countries have increased their reserves substantially. Given the fact that exploration costs in these countries are low compared to other countries, the potential to increase production capacity is large.

Venezuela) favour relatively low oil prices to keep marginal producers out of the market. Their production potential is, after major recent discoveries, very large. Their national oil companies have also embarked on vertical integration, and they have begun to behave like multinationals. Depending on the degree of their 'freedom' from government interference, these companies will become increasingly vertically integrated. This requires a different approach to the market. Already, these countries insist on higher production levels than the quota allotments in order to supply their refineries. This change in market approach can become more pronounced when state oil companies increase their operations in consumer markets, either through acquisitions or joint-ventures.

Following the price collapse of 1985/1986, the oil market is once more in a state of transition. The power of the cartel has become diffused. Oil exploration decreased due to the price collapse, demand recovered, and refinery margins were improved. The cycle leading to a tight oil market was again set in motion. The dominant players in the next cycle of tight supply, compared to demand, will certainly include the surplus countries and the majors. They will most probably end up at one side of the market, with the OPEC price hawks on the other side.

Chapter Seven

Market Process and Structure

1. Market cycles

The inherent instability in the oil industry is partly caused by counter-cyclical conditions in the separate stages of production, but also by high fixed costs, technical constraints, the variety of costs between wells, incomplete homogeneity of crude oil, and the inflexibility generated by joint production in the processing phase. Supply and demand, although quite inelastic in the short-term, have proven to be more elastic in the long-term. These characteristics have proved advantageous to cartel formation in the short-run, but a limitation on their durability in the long-run.

Investments in the oil industry tend to have a pro-cyclical effect, which is partly caused by the relatively long time-lag between the investment decision and actual production, the uncertainty about the actual size of a well, but also by the dubious record of forecasting supply and demand on which these decisions are made. Current oil prices, the utilization rate of capacity in place, and supply and demand expectations, have a strong influence on investments. The degree of (anticipated) government intervention in the industry is an additional consideration in investment decisions.

Exploration involves large investments without certainty about the quantity and quality of oil that can be recovered. In general, high current oil prices as compared to costs have stimulated oil exploration, and low oil prices have delayed exploration. Since finding and developing a field takes time, oil might enter the market under rather different economic conditions. Still, once production capacity is in place, oil will remain on stream from that well as long as it covers the low operating costs, thus increasing the price volatility. A period of tight supply pushes prices up, and exploration increases. With an interval, demand reacts to higher oil prices and slackens. On the supply side, the increase in exploratory activity results in time in new additions to production, and with demand decelerating, the market relaxes further and prices decline. A surplus is thus predetermined in a tight market. Conversely, in a buyers' market, investments in upstream capacity extensions are delayed and will eventually result in a sellers' market and price increases.

Oil prices and demand developments influence capacity adjustments in the downstream industry in a reverse manner. The divergent (in)elasticities in the product markets, the fact that these products are derived in joint-production, and the capital intensity of production inhibit smooth adjustment to alternating market developments. The market generates capacity-bottlenecks (surplus or deficit) at, and between, the various stages of production. The inability of the market participants to prevent these leads from one shock to another. These shocks reverberate through the industry and require continuous adjustments.

Although the upstream and downstream developments are counter-cyclical, they are not each others' perfect mirror image. This is the result of the unequal elasticities of supply and demand in the various stages of production. Thus, the upstream and downstream cyclical fluctuations tend partly to emphasise the fluctuations of the oil market cycle, rather than compensating each other. The organisational structure of the market participants is moulded by these market fluctuations.

In particular, the fundamental counter-cyclical development of the upstream and downstream sectors and the high capital intensity of exploration and processing have been incentives for vertical integration of companies, enabling the companies to internalise the fluctuations of the market cycle. This process of internalization reduced their investment risks and made it possible to counterbalance profits and losses in the different stages of production, reducing taxation, stabilising dividend payments, and otherwise serving their long-term interests.

Other factors influencing forward and backward integration have been geological, geographic, technical and infrastructural constraints. Furthermore, the incomplete, or absence of, flexible markets placed a premium on security of crude supply for producers and processors alike. It gave refiners the desired security of supply, and producers of oil a more secure market outlet.[1] A high self-sufficiency ratio of oil production over processing was considered a sound business strategy by the companies. Once the large international companies had developed into vertically integrated firms, trading outside company channels became marginal.[2] Only when OPEC drove the companies into backward disintegration at the end of the 1970s, at a time when the OPEC countries themselves still lacked the processing and distribution channels, did more (short-term) trading develop.[3] Within a short time, trade in oil was revolutionised. A great variety of oil markets emerged. These spot, forward and future markets not only served 'wet', but also 'paper' trade, and assisted to hedge trading risks.

The loss of equity oil motivated the companies to begin immediately a process of backward reintegration. The price policy of OPEC, stabilising oil prices at a high level, promoted exploration activities, while the stockmarket value of some oil companies stimulated takeovers. Thus the companies engaged in a mix of exploration activities (outside

[1] See Chapter two and four.

[2] See Chapter four.

[3] See Chapter five and six.

the OPEC countries) and mergers and acquisitions to increase their equity crude production.[4] The companies appear to be reinstalling their high self-sufficiency in crude oil production over processing. Consequently, the self-sufficiency ratio of the largest 12 international oil companies was at its lowest in 1981, at only 48%. This level increased to 63% in 1985/1986, but has slipped back to 60% in 1989.[5]

The improvement of the self-sufficiency ratio of the companies slipped when exploratory spending was reduced after the 1986 oil price collapse. The fact that their fields in the US and the North Sea have, or are believed to have, passed their peak production levels, and that 'Wall Street' has run dry, also affected their ratio. The fall of oil prices in the second half of the 1980s, and its effects on exploration, increased the concern that delayed investments in new production capacity will prompt another price shock in the near future when capacity utilisation is increasing, thus proving again the inability of industry to escape the flux of the market cycle.

The flow of the market cycle also manifested itself in the downstream industry. Until the 1970s, the international oil companies were still mainly crude oil producing entities that had moved upmarket in processing and in some cases integrated horizontally in other energy producing activities. Their processing capacity was basically derived from their crude production capacity, and imbalances between production and processing were compensated by long-term contracts. The companies' profits were secured by increased turn-over of crude oil and sales of their main oil products. The other refining products were long considered 'by-products'. However, the economics of joint-production demanded expansion of these additional product markets, when refinery output continued to increase. The production volume of these 'other' oil products was predominantly correlated to the required output of the predominant oil products, not to demand in these particular product markets. Treating these products as 'by-products' became increasingly costly to the industry.

The relatively low oil prices in the 1960s kept this predicament of the processing industry temporarily hidden, because refinery margins improved. Low oil prices decreased the value of the 'by-products', which facilitated competitive pricing with regard to their near substitutes. However, at the end of the 1960s, capacity bottlenecks were becoming evident in both the upstream and the downstream sectors. The utilisation of crude production capacity was increasing, and an upturn in the price trend was forthcoming. An increase in crude oil prices would bolster exploration again, solving the upstream capacity problems. However, higher oil prices could not solve the downstream capacity bottlenecks.

The impending maturity of their main consumer markets led the majors increasingly to question their 'quantitative' strategy. In the early 1970s, it had become evident that they needed to restructure their operations, but they found themselves caught in the counter-cyclical nature of their operations to make the necessary adjustments.

4 See Chapter six.

5 *Petroleum Intelligence Weekly*, June 11, 1990.

The price increases of the early 1970s, initiated by producer governments seeking an increase in their revenue, unlocked this impasse, and capacity adjustments in the upstream and downstream sector were made. The processing industry began to improve refinery flexibility, to upgrade its plants, to diversify its markets, and to implement better cost management. In general, refinery capacity was diminished.[6] The former 'quantitative' approach was replaced by a more qualitative one, and in many companies separate divisions were designated as individual profit-centres.

The downstream industry in the OECD countries was adapted to meet the quantitative and qualitative changes in demand in these mature consumer markets. For the past few years, demand continued to expand to the lighter end of the barrel.[7] This qualitative change in demand resulted in either very small rates of growth, or even in a decrease in total oil demand in the OECD countries.[8] The development of demand in developing countries is almost the opposite. Despite the price increases, demand for light as well as heavier oil products has continued to grow. In 1989, oil consumption in some Asian countries exceeded 10%, an indication that these markets are far from mature.[9]

One possible present downstream capacity bottleneck results from the fact that processing capacity has been geared too much towards changing conditions in the OECD markets, not taking the expansion of the less mature markets into account. Further, the impending implementation of environmental legislation in the OECD countries to restrict emission levels of the processing industry itself, and to force them to produce 'cleaner' products, might aggravate the short-term capacity bottlenecks. The processing industry will be forced to invest in cleaning up their operations, thus stressing the qualitative aspect of capacity, while growing demand from other regions will require capacity extensions.

2. Growth cycles

The oil industry life cycle can be divided into three distinct industry growth cycles, each of about forty years. These are characterized by long and substantial periods of expansion (about twenty years), and a comparatively short period of maturity before going into stagnation or decline. The latter two phases exhibited an increasing degree of concentration and cartel formation. The combined duration of these two phases is also approximately twenty years. The oil cycle was repeated when new markets (in terms of products and in terms of consumer markets) were developed, although the range of variation in average annual growth rates became smaller in each subsequent cycle (See Figure 7.1). The narrowing rates of growth signify the growing maturity of the oil industry, and its increasing

[6] See Chapters one and six.

[7] See Chapter one.

[8] See Chapter three.

[9] *BPSR 1990.*

penetration in the economy as a whole. The declining rate of substitution (or even the occurrence of inverse substitution), the declining energy/GNP and oil/GNP ratio, and a shift towards more value added oil products, characterize this development. The US market was the first to mature. The US post-war average annual growth rates of consumption have been substantially lower than those of the international market. A similar development is to be expected for the other major consumer markets, Western Europe and Japan.

Between 1860 and 1988, the oil industry completed three of these long growth cycles. The lamp oil cycle lasted from 1860 to about 1900. The fuel cycle can be broken into two distinct long cycles: the American fuel cycle (1900-1940), and the international fuel cycle (1950-).

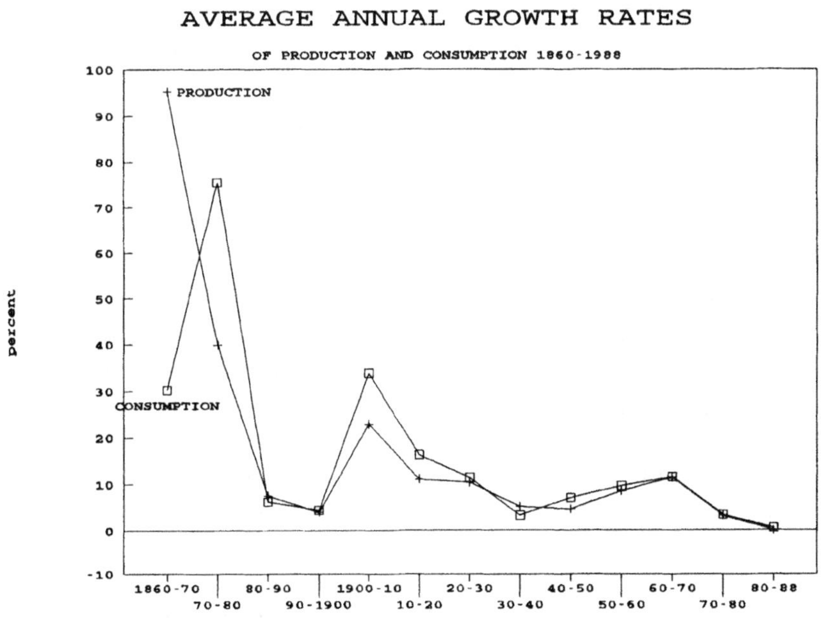

Figure 7.1

In the course of the industry life cycle certain oil products 'dropped out', i.e., they entered a stagnation phase in certain consumer markets, while alternative or new oil products were in, or were about to enter, an expansion phase. The oil industry was at first predominantly a lamp oil industry. At the turn of the century, it progressed into a fuel producing industry, with gasolines and gas and diesel oils as its main products. In the 1970s, the international gas and diesel oil markets stagnated, when demand also shifted to the lighter end of the barrel in Western Europe and Japan. The gasoline cycle merely temporarily flattened out, due to price increases. When prices declined, demand for gasoline recuperated because motor fuels remain costly to substitute. Nevertheless, the international fuel cycle has been maturing. Although the energy use of oil is still predominant, there has been a distinct shift towards greater non-energy use. The latter trend will continue, thus strengthening the link between refining and the petrochemical industry.[10] The next growth cycle is certain to be dominated by high value-added fuels and non-energy oil products in the mature consumer markets, and heavier oil products in the expanding markets of developing countries.

These changes in the use of oil brought about a repetition of phases of production in the industry growth cycles and also a repetition of changes in market structures. The transition periods between the major growth cycles (approximately 1895-1905, 1930-1940, and 1980-1990) are characterized by an overlap in time of the 'use' of oil, an unsynchronized maturity of consumer markets and of specific oil product markets, and upstream and downstream capacity bottlenecks. In addition to the 'major' transition periods, which mark the transition to another long growth cycle, transitions within a growth cycle (from one phase of production to another) have been identified. The start of a new long growth cycle is marked by rapid expansion and competitive markets, while the progression along the growth cycle shows increasing maturity and finally stagnation. Maturing markets have exhibited decreasing rates of growth and increased concentration and cartel formation.

The transitional periods connect the shifting market structures, when a process of deconcentration or concentration increases but is not yet dominant. In periods of emerging concentration, merger and take-over activity increased, while in periods of deconcentration new entrants entered the market. These developments coincided with an upturn in the trend of average annual growth rates of consumption and production. The identification of an upturn or downturn in the market trend is crucial, since it makes it possible to draw a distinction between (shorter) market cycles and a new phase of production in the growth cycle. The latter will result in a new market structure, the former in a resumption of the previous trend.

In order to identify a change in the market trend, the (average annual) rate of growth of production and consumption, and the level of integration of market participants, must be supplemented by the long-term development of (real) prices. Prices were relatively high both in nominal and real terms in the periods 1860-1870; 1895-1900; 1916-1926; 1945-1960, and 1973-1982, while they dropped conspicuously in the periods 1875-1880; 1900-1905; 1925-1930; 1965-1970; and 1982-1988 (See Figure 2.1). These periods of changing price trends were an expression of changing market conditions and the subsequent capacity bottlenecks (deficits or surpluses) between the upstream and the downstream sector. The price fluctuations in the periods of transition (initial relative high nominal prices followed by lower prices) were a result of the difficulties the industry encountered in making smooth

[10] For instance, implementation of environmental legislation intended to reduce gasoline fumes requires the use of petrochemicals to produce cleaner gasolines, while the petrochemical industry obtains feedstock from the refining industry.

adjustments. This is due to a difference in time-lags in capacity adjustments in the upstream and downstream sectors.[11] Depending on the evolution of price effects, an upward price trend set a new round of capacity extentions in motion, while a downward trend generally slowed down exploration efforts. This ongoing flow of oil market cycle features throughout the modern history of the industry.

As with the declining rate of change in average annual growth rates, price fluctuations also decreased. This trend is only broken after World Wars I and II (See Figure 2.1), due to war-time inflation, and in 1973/1974 and 1979/80. The oil price increases of 1973/1974 and 1979/1980 must be seen **in part** as political inflation.[12] The existing upward price trend in the early 1970s was magnified as a result of an ongoing 'political struggle' over the redefinition of the international economic order, i.e., the struggle between industrialized and developing countries to capture economic rents.

In this study, three distinct phases of expansion were identified: 1860-1880, expansion of the lamp oil market (See Chapter two); 1900-1930, expansion of the US fuel market (See Chapter two); and 1950-1973, expansion of the international fuel market (See Chapter three). These periods were characterized by increasing deconcentration and a competitive market.

The maturity phase of each period was characterized by a high degree of concentration and cartels. These phases were: 1880-1900, Standard Oil monopoly (See Chapter two); 1930-1940; the International Petroleum Cartel (See Chapters two and four); and 1973-1985, OPEC (See Chapters three, five and six). These periods of expansion and maturity were classified according to the average annual growth rates of production and consumption presented in Chapters two and three, rates which were graphically displayed in Figure 7.1.

3. Market structure

Dynamic market theory contends that in the process of a growth cycle, three essential moments with respect to concentration and deconcentration and, in a broader sense, relating to changes in the market structure, would be visible. In the introductory phase horizontal and vertical integration is said to occur. A process of deconcentration (both vertical and horizontal) would then take place in the transition from the introduction phase to the expansion phase and during a large part of the latter. Thereafter, concentration (vertical and horizontal integration) would gain prominence at the end of the expansion phase and at the beginning of the maturity phase, as well as during the latter. In the maturity phase, the probability of collusive agreements between companies increases, as price competition becomes an unattractive option. In this stage, the number of mergers and takeovers also increases.

The counter-cyclical conditions in the upstream and downstream sectors have been a

[11] See section 1.

[12] See section 4.

continuous incentive for vertical integration within the oil industry. However, the level of integration of the companies compared to the market varied in the different stages of the growth cycle. Similar patterns were also found in the lamp oil cycle and the fuel cycles. Concentration was tangible when vertical and horizontal integration became stronger, and when collusion and cartel formation occurred, while deconcentration was notable when the number of companies entering the (international) market increased.

The identification of three complete industry growth cycles, substantiated by the identification of the periods of expansion, maturity and decline, should, according to theory, be mirrored in the oscillations of market structures, and thus in the level of concentration.

Several tools have been developed to determine the level of concentration in a market. The pursuit of the objective to determine exactly market structures, and to measure them mathematically, resulted in the development, among others indicators of concentration and competition, of the concept of the concentration ratio. However, on its own, this measurement of market concentration or deconcentration presents a rather static picture of the market, and it fails to recognise a change in the trend. In terms of concentration ratios the US market, for instance, was still quite concentrated between 1900 and 1910 (See Chapter two). The impact of the new entrants, who quickly captured a market share in the relatively small but rapidly expanding fuel market, could not be measured adequately. Deconcentration was latent. The same is true for the period that the International Petroleum Cartel was in control. Its heyday was between 1930 and 1940. During the Second World War and its aftermath, a status quo prevailed. However, from 1950 onwards new entrants entered the international oil market. Although the concentration ratios are still substantial in this period, the power of the cartel had already diminished (See Chapter four).

The analysis of the long sweep of international oil market developments and the logic of changing market structures was carried out primarily to arrive at an understanding of the operation of the industry. The hypothesis, developed in Chapter one, that the oil industry encompassed a stratification of growth cycles, suggested some of the possible forces and logic at work, but only in an abstract way. This resulted in a rather descriptive approach. In the course of this study, the focus on crude oil only, and the omission of consistent measurement of actual levels of concentration or deconcentration in the upstream and the downstream sectors, presuming that the relevant data could have been obtained for the entire period, revealed certain drawbacks. The fact that the concentration ratios cited from various sources in Chapters two and four disguised the periods of latent concentration and deconcentration only emerged when the findings on market developments and structures was put together. However, these periods must now be considered vital to explain the logic of changing market structures. The findings were that these periods coincided with a change in the average annual growth rates, and that price trends changed. However, at this point, the relationship between the evolution of the growth cycle, the oil product cycles, and the accompanying changing market structures in these particular periods, as argued by the dynamic market theory, can be substantiated by circumstantial evidence or by logical reasoning. Although the identification of the periods of expansion, maturity and decline have sharpened analysis of these (market structure) transition periods, substantive additional research is required. In particular, the periods of latent concentration or deconcentration must be studied in greater detail and must also include the evolution of product markets. A comparison of output growth in (new) product markets and innovations in the production process, of established firms and new entrants, could throw more light on the impact of the latter on the vested market position of established firms.

This study therefore reveals important issues for further investigation, including the connection between the flow of the industry growth cycle with the changing market structures during transitional periods. At this point it will be possible to establish periods of concentration and deconcentration, which coincide with expanding and maturing and declining markets. The evidence which allowed us to link the periods of expansion and maturity and decline to competition and cartelisation of the market was presented in the previous chapters. Between 1860 and 1880, 1900 and 1920, and 1950 and 1970, the periods of expansion were also the periods in which the number of (non-integrated) market participants increased substantially. Between 1880 and 1900, 1930 and 1940, and 1970 and 1985, the periods of maturity and decline, great market concentration (mergers and acquisitions) and cartel formation was observed.

In the transition periods, the market cycle revealed capacity bottlenecks that manifested itself in changing trend of annual average growth rates (See Figure 7.1) and prices. Another feature of these periods was the intensification of government intervention. A change in the market structure is assumed to be latent in these periods: 1870-1880; 1895-1905; 1925-1930; 1950-1960; 1970-1973; 1982-1985.

4. Government intervention

In the course of the modern history of the oil industry, government intervention ranged from regulation of exploration, production, processing, pricing, and demand, to actual participation in the market through state companies. Taxation has played a major part in the development of markets (See Chapter four). The internationalization of the oil industry meant that government intervention increasingly became a mix of domestic market regulation and trade regulation. Moreover, internationalization sharpened the conflict of interests between governments and industry strategies. In general, government intervention intensified over time.

In the context of this long sweep of oil market developments, one may question the impact of government intervention on the flow of growth cycles. Are governments better equipped or able to overcome the inherent instability of the industry, and prevent capacity bottlenecks between the upstream and downstream industry? Or are governments in the long-run obliged to follow the market developments, like other market participants?

In many cases, government intervention turned out to be pro-cyclical, and merely hastened developments already taking place. This pro-cyclical effect is both due to a delay in the actual implementation of measures and in its results. Government measures tend to follow market cycles rather than anticipate them, and measures regularly only take effect under already different market conditions. Moreover, measures seem to survive longer than the specific market conditions they were meant to regulate. Examples of pro-cyclical government intervention were: concession policies of newly-producing countries in the 1960s, the US import quota scheme, the invocation of the OECD oil stock policy in 1979, OPEC's price policy in the early 1980s, US consumer price protection, etc.

The impact of government intervention on market conditions has been considerable. The findings were, however, that government intervention, which aimed to turn, to obstruct, to dismiss or to promote certain market trends is similar to attempts by other market participants, i.e. it is limited in effect and duration. Despite its limited scope, the desire of governments to intervene in the market is substantial, since oil is an important input in the (international) economy. Producer governments stand to gain part of the economic rents, and to extend economic and political control beyond their national frontiers. Consumer governments want to regulate their dependence on imports, and they have at the same time been quite adept in raising income from taxes on oil products. Further study of government intervention in the context of the dynamic market theory is definitely required.

Government intervention has increased throughout the years. It developed from applying anti-trust legislation, to actual participation in the industry as producers and processors, and has developed from domestic regulation to international regulation. The extent of government regulation is related to the importance of oil to the (international) economy, the absence of domestic firms in the industry, and the increasing maturity of major markets. In this respect, OPEC was a culmination of growing government intervention.

Consumer government interference with the oil industry is unlikely to diminish in the future. Although President Reagan decided to de-regulate the industry in the early 1980s, and 'let the market regulate the industry', the pressure on the Bush Administration to define a new energy policy is becoming stronger. In Spring 1990, the US imported more than 50% of its domestic consumption. Low oil prices had stimulated demand, and energy conservation policies seemed to belong to the past. The US is now facing a slow but continuing decline of its oil production. New exploration meets increasing opposition from the environmental lobby. Especially after the Alaska oil spill in 1989, several projects were mothballed. For the time being, import dependency is unlikely to diminish. Oil is already responsible for a large part of the trade deficit, and since this must be considered a structural import, the economic problems of the government are mounting. The Bush Administration admitted in Spring 1990 that the US was forced to rely upon the international exploration efforts of the oil companies to satisfy demand.

In Western Europe, governments continued to intervene in the oil market. Oil price decreases were not completely passed on to consumers to inhibit an unwanted expansion of demand. The present tax burden on oil products is not likely to decrease, and will only increase when 'green taxes' are introduced. Taxes also protect the consumers from steep price increases, since the crude oil price is only a part of the consumer price. In the US, taxes on oil products are much lower, and prices have a much larger effect on consumption. However, debates on the introduction of 'green taxes' leading to an increase in gasoline prices have met stubborn resistance. In Western Europe and Japan, consumers are already used to higher product prices. The North Sea production has been a source of secure supply for West European countries. The fact that supply is expected to decrease slowly will stimulate intra-European oil trade. The dependency on OPEC oil is likely to increase again.

Thus far, government intervention in the oil consuming developing countries has been limited. The developing countries steadily increased oil consumption in the 1970s and 1980s. This expansion of demand will continue for some time to come. Between 1986 and 1990, annual growth of demand in some Asian countries exceeded 10%, fuelling their rapidly growing economies.

The impact of OPEC on the world market has diminished in the last ten years, and the organisation is suffering from the diverging interests of its the member states. Its policies have become less unified, and at times member states pursued policies that actually hurt other member states. This has resulted in less consistent market behaviour. The period 1985-1990 has been more dominated by the large net exporters, pursuing a low price policy with the backing of some other member states. However, the alliances among member states over policy issues have changed. This is in part due to political reasons and in part due to economic reasons. The Iran-Iraq war certainly had its influence on these changing alliances. Government intervention in OPEC member states will remain significant, although some member states will increasingly pursue their policies through their (more independently operating) state oil companies. How that will effect OPEC remains to be seen, but most certainly OPEC has entered a new phase in its existence.

Bibliography

Aoki, M., Gustafsson, B., Williamson, O.E. (eds.), *The Firm as a Nexus of Treaties*, Sage Publications, London, 1990.

Adelman, M.A., OPEC as a cartel, in: *OPEC Behavior and World Oil Prices*, J.M. Griffin and D.J. Teece (eds.), George Allen & Unwin, London,1982.

Adelman, M.A., *The World Petroleum Market*, Johns Hopkins University Press, Baltimore, 1972.

Aglietta,M., *A theory of Capitalist regulation; the US experience*, NLB, London, 1979.

Al-Chalabi, F.J., OPEC today: OPEC tomorrow, in: *OPEC Review*, vol. IX, no. 3, Autumn 1985, p.211-221.

Al-Chalabi, F.J., OPEC and the present structural limitations on its oil price control, in: *OPEC Review*, vol. XII, no. 2, Summer 1988, p.115-121.

Al-Chalabi, F.J., The world oil market in the 1990s, in: *OPEC Review*, vol. XII, no. 3, Autumn 1988, p.227-238.

Al-Chalabi, F., *OPEC and the International Oil Industry, a changing structure*, Oxford University Press/OAPEC, Oxford, 1980.

Al-Fathi, S.A., OPEC's refining industry- the developments and prospects, in: *OPEC Review*, vol. XIII, no. 2, Summer 1989, p.91-110.

Alnasrawi, A., *OPEC in a Changing World Economy*, The Johns Hopkins University Press, Baltimore/London, 1985.

Al-Otaiba, M.S., *OPEC and the Petroleum Industry*, Croom Helm, London, 1975.

Al-Yousuf, A., *Kuwait & Saudi Arabia: From Prosperity to Retrenchment*, Oxford Institute for Energy Studies, Oxford, 1990.

American Petroleum Institute, *Petroleum Facts and Figures*, New York, 1937.

American Petroleum Institute, *Petroleum Facts and Figures*, Washington D.C., 1971 edition.

American Petroleum Institute, *Basic Petroleum Data Book, Petroleum Industry Statistics*, Washinton D.C., 1984-1990.

Archer, L., *The Gulf Crisis: Implications for the Environment*, GWO8, Oxford Institute for Energy Studies, Oxford, 1990.

Attiga, A.A., *Interdependence on the Oil Bridge, Risks and Opportunities*, Petroleum Information Committee of the Arab Gulf States, 1988.

Auerbach, P., *Competition, The Economics of Industrial Change*, Basil Blackwell, Oxford, 1988.

Bacon, R., Chadwick, M., Dargay, J., Long, D., Mabro, R., *Demand, Prices and the Refining Industry*, Oxford Institute for Energy Studies/ Oxford University Press, 1990.

Bain, J.S., *Industrial Organization*, John Wiley and Sons, New York, 1959.

Ballance, R.H., Sinclair, S.W., *Collapse and Survival: Industry Strategies in a Changing World*, George Allen & Unwin, Boston/Sydney, 1983.

Ballance, R.B., *International Industry and Business, Structural Change, Industrial Policy and Industry Strategies*, Allen & Unwin, Boston/Sydney/Wellington, 1987.

Barnikel, H-H., *Theorie und Praxis der Kartelle*, Wissenschaftliche Buchgesellschaft, Darmstadt, 1972.

Barnes, P., *Can Natural Gas Take the Strain*, GWO5, Oxford Institute for Energy Studies, Oxford, 1990.

Beredjick, N., Walde, T., *Petroleum Investment Policies in Developing Countries*, Graham & Trotman, London/Dordrecht/Boston, 1988.

Berry, B.J.L., *Long-wave Rhythms in Economic Development and Political Behavior*, The Johns Hopkins University Press, Baltimore, 1991.

Best, M.H., *The New Competition, Institutions of Industrial Restructuring*, Polity Press, Cambridge, 1990.

Betkenhagen, J., The impact of energy on East-West trade: retrospects and prospects, in: *East-West Trade and Finance in the World Economy, a new look for the 1980s*, Saunders (ed.), C.T., Macmillan, London, 1985.

Biekart, J., Steenkool en Macht, in: *Intermediair*, 1 februari 1985.
Bird, G., *World Finance and Adjustment, an agenda for reform*, Macmillan, Basingstoke and London, 1985.
Black, F., *Business Cycles and Equilibrium*, Basil Blackwell, Oxford, 1990.
Blair, J.M., *The Control of Oil*, Vintage Books, New York, 1978.
Baumol, W.J., Blinder, A.S., *Economics, Principles and Policy*, Harcourt Brace Jovanovich, San Diego/New York, 1991.
Boers, J., Eisenloeffel, G., Linde, C.van der, Olieraffinage en petrochemie in de Golfstaten en EG, in: *Economisch Statistische Berichten*, 22-2-1984.
Boot, P., Economische stagnatie in Oost-Europa, in: *Internationale Spectator*, oktober 1983, xxxvii-10.
BP, *British Petroleum Statistical Review of the world oil industry*, 1949-1990.
Bronner, W., Falk, R., e.o., *Ölkrise und arabische öllander*, Marxixmus aktuell, Frankfurt am Main, 1980.
Brown, W.M., *Petroleum Prices: Past, Present, and Prospective*, Hudson Institute, Indianapolis, 1987.

Caffarra, C., *The Role and Behaviour of Oil Inventories*, GWO4, Oxford Institute for Energy Studies, Oxford, 1990.
Chadwick, M., Long, D., Nissanke, M., *Soviet Oil Exports, trade adjustments, refining constraints and market behaviour*, O.I.E.S./Oxford University Press, Oxford, 1987.
Chase Manhattan Bank, *Capital Investments of the World Petroleum Industry*, 1964-1984, New York.
Chase Manhattan, *Financial Analysis of a Group of Petroleum Companies*, 1968-1985, New York.
Clark, R., McGuinness, T., *The Economics of the Firm*, Basil Blackwell, Oxford, 1987.
Coase, R.H., *The Firm, the Market, and the Law*, University of Chicago Press, Chicago/London, 1988.
Cox, H., Jens, U., Markert, K., *Handbuch des Wettbewerbs*, Verlag Franz Vahlen, Munchen, 1981.

Daintith, T., Hancher, L., *Energy Strategy in Europe: The Legal Framework*, series A, Law, volume 4, Walter de Gruyter, Berlin/New York, 1986.
Daintith, T., Williams, S.F., *The Legal Integration of Energy Markets*, series Integration through Law, volume 5, Walter de Gruyter, Berlin/New York, 1987.
Danielsen, A.L., *The Evolution of OPEC*, Harcourt Brace Jovanovich, Publishers, New York, 1982.
Daoudi, M.S., *The politactics of international cartels: economic illusions, political, realities, and OPEC*, University Microfilms International, Ann Arbor, 1985.
Dargay, J., *Oil Demand: Dependence or Flexibility?*, GWO6, Oxford Institute for Energy Studies, Oxford, 1990.
Darmstadter, J., with Teitelbaum, P.D., and Polach, J.G., *Energy in the World Economy, a statistical review of trends in output, trade, consumption since 1925*, Johns Hopkins University Press, Baltimore/London, 1971.
Davidson, K.M., *Mega-mergers, Corporate America's billion dollar takeovers*, Ballanger Publishing Company, Cambridge, Mass., 1985.
Davis, J.D., *Blue Gold: The political economy of natural gas*, George Allen & Unwin, London, 1984.
Davison, A., Hurst, C., Mabro, R., *Natural Gas: Governments and Oil Companies in the Third World*, O.I.E.S./ Oxford University Press, Oxford, 1988.
Demsetz, H., *Ownership Control and the Firm, The Organization of Economic Activity Volume I*, Basil Blackwell, Oxford, 1988.
Demsetz, H., *Efficiency Competition and Policy, The Organization of Economic Activity Volume II*, Basil Blackwell, Oxford, 1989.
Dunning, J.H., *The Multinational Enterprise*, George Allen & Unwin, London, 1974.
Dunning, J.H., *Explaining International Production*, Unwin Hyman, London, 1988.
Duyn van, J.J., *De lange Golf in de Economie, Kan innovatie ons uit het dal helpen?*, Van Gorcum, Assen, 1979.

Eatwell, J., Milgate, M., Newman, P. (eds.), *The Invisible Hand*, The New Palgrave, The Macmillan Press Ltd., 1989.
Eatwell, J., Milgate, M., Newman, P. (eds.), *Problems of the Planned Economy*, The New Palgrave, The Macmillan Press Ltd., 1989.
Eatwell, J., Milgate, M., Newman, P. (eds.), *Capital Theory*, The New Palgrave, The Macmillan Press Ltd., 1989.

Edwards, C., *The Fragmented World, Competing Perspectives on Trade, Money and Crisis*, Methuen, London/New York, 1985.
Eggertsson, T., *Economic behavior and institutions*, Cambridge University Press, Cambridge, 1990.
Ehrlich, E.M., Structural Change in OPEC Member Country Production Decisions, in: *OPEC Review*, vol. IV, no. 4, Winter 1980, p.97-113.
El-Imam, M.M., Decline in Arab Oil Revenues, in: *The Decline of Arab Oil Revenues*, A.M. Farid and H. Sirriyeh (eds), Croom Helm, London, 1986.
El-Mallakh, R., *Heavy versus Light Oil: Technical and economic considerations*, ICEED, Boulder, Colorado, 1983.
El-Mallakh (ed.), R., *OPEC: Twenty Years and Beyond*, Westview Press/Croom Helm, Colorado/London, 1982.
El-Mangosh, A., Refining flexibility in upgrading heavy oil, in: *OPEC Bulletin*, vol. XIII, no. 2,, March 1982, p.1-12.
Emery, J.J., Graham, N.A., Oppenheimer, M.F., *Technology Trade with the Middle East*, Westview Press, Boulder/London, 1986.
Evans, J., *OPEC, its member states and the world energy market*, Keesing's reference publication, Longman, Harlow, 1986.

Fanning, L.M., *The Story of the Petroleum Institute*, API, Washington, D.C., 1959.
Farid, A.M., Sirriyeh, H. (eds.), *The Decline of Arab Oil Revenues*, Croom Helm/Arab Research Centre, London, 1986.
Farrell, T.M.A., The world oil market 1973-1983, and the future of oil prices, in : *OPEC Review*, vol. IX, no. 4, Winter 1985, p.389-416.
Ferrier, R.W., *The History of the British Petroleum Company, volume 1; the developing years 1901-1932*, Cambridge University Press, Cambridge, 1982.
Finlayson, J.A., Zacher, M.W., *Managing International Markets, developing countries and the commodity trade regime*, Columbia University Press, New York, 1988.
Fisher, D., *North Sea Oil, an environment interface*, Universitetsforlaget, Bergen/Oslo/Tromso, 1981.
Friedrich Ebert Stiftung, *Internationale Politik, Saudi-Arabien in den 80er Jahren*, Expertengesprach 26. und 27. April 1982.
Fröbel, F., Heinrichs, J., Kreye, O., *Die neue internationale Arbeitsteilung*, RORORO, Reinbek bei Hamburg, 1977.

Gerretson, C., *Geschiedenis der 'Koninklijke'*, 3 volumes, Oosthoek, Utrecht, 1939/1942.
Gilpin, R., *The Political Economy of International Relations*, Princeton University Press, Princeton, 1987.
Goodwin, C.D., Energy: 1945-1980, in: *The Wilson Quarterly*, Spring 1981.
Grayson, L.E., *National Oil Companies*, John Wiley and Sons, Chichester/New York, 1981.
Grayson, L.E., *Who and How in Planning for Large Companies, Generalizations from the Experiences of Oil Companies*, Macmillan, Basingstoke/London, 1987.
Griffin, J.M., Teece, D.J., *OPEC Behavior and World Oil Prices*, George Allen & Unwin, Boston/Sydney, 1982.
Gustafson, T., *Crisis amid Plenty, The Politics of Soviet Energy under Brezhnev and Gorbachev*, Princeton University Press, Princeton, 1989.

Haan, W., Molen van der, F., Linde, C. van der, Een zwaargewicht slankt af: Unilever 1965-1980, in: *Nederlands Kapitaal over de Grenzen*, F. Crone and H. Overbeek (eds.), SUA, Amsterdam, 1981.
Hamilton, A., *Oil, The Price of Power*, Michael Joseph/Rainbird, London, 1986.
Hammer, A., Lyndon, N., *Hammer, witness to history*, Coronet Books, Sevenoaks, 1987.
Hawdon, D., *The Energy Crisis, ten years after*, Croom Helm/St. Martin's Press, London/New York, 1984.
Hawdon, D., *The Changing Nature of the World Oil Industry*, Croom Helm, London/Sydney, 1985.
Hawdon (ed.), D., *Oil Prices in the 1990s*, Macmillan, Basingstoke/London, 1989.
Helm, D. (ed.), *The Economic Borders of the State*, Oxford University Press, Oxford, 1990.
Hendriks (ed.), F., *Shell, produktie, verwerking, verkoop, energie, metalen, bosbouw, zaaizaad, casestudies*, SOMO/Uitgeverij Jan van Arkel, Amsterdam/Utrecht, 1987.

Hickman, B.G., *Growth and Stability of the Postwar Economy*, The Brookings Institution, Washington D.C., 1960.

Hoffman, G.W., *The European Energy Challenge; East and West*, Duke Press Policy Studies, Durham N.C., 1985.

Holden, D., Jones, R., *The House of Saud*, Pan Books, London, 1982.

Hood, N., Young, S., *The Economics of Multinational Enterprise*, Longman, Harlow, 1979.

Horsnell, P., *Oil Price Differentials: Markets in Disarray*, GWO3, Oxford Institute for Energy Studies, Oxford, 1990.

Hussein, A., *Iraq: The Eternal Fire, 1972 Iraqi oil nationalization in perspective*, Third World Centre for Research and Publishing, London, 1981.

Idemudia, T., OPEC and the international oil market: continuity and change in the 1990s, in: *OPEC Review*, vol. XIII, no. 2, Summer 1989, p.111-131.

IEA, *Energy Balances of OECD Countries 1987-1988*, OECD, Paris, 1990.

IMF, *World Economic Outlook, a survey by the Staff of the International Monetary Fund*, Washington, D.C., various issues.

Jacoby, N.H., *Multinational Oil, A Study in Industrial Dynamics*, Macmillan Publishing Co./Collier Macmillan Publishers, New York/London, 1974.

Jacquemin, A., *The New Industrial Organization, Market Forces and Strategic Behavior*, Clarendon Press, Oxford, 1987.

Jenkins, G., The uncertainties in oil products forecasting to the year 2000, in: *OPEC Review*, vol. XII, no. 2, Summer 1988, p.145-164.

Johany, A.D., *The myth of the OPEC cartel, the role of Saudi Arabia*, John Wiley & Sons, Chichester, 1982.

Jones, H., *An introduction to Modern Theories of Economic Growth*, Nelson Walton-on-Thames, 1981.

Jong de, H.W., *Dynamische Markttheorie*, Stenfert Kroese, Leiden, 1985.

Jong de, H.W., *Dynamische Concentratietheorie*, Stenfert Kroese, Leiden, 1972.

Jong de (ed.), H.W., *The Structure of European Industry*, Kluwer Academic Publishers, Dordrecht/Boston/London, 1988.

Jong de, H.W., Shepherd, W.G., *Mainstreams in Industrial Organization*, 2 volumes, Kluwer Academic Publishers, Dordrecht, 1986.

Julius, D., *Global Companies and Public Policy, The growing Challenge of Foreign Direct Investment*, The Royal Institute of International Affars/ Pinter Publishers, London, 1990.

Keegan, W., *Britain without Oil*, Penguin Books, Harmondsworth, 1985.

Kahn, A.R., The Prospects for Gas in Developing Countries, in: *Natural Gas, an international perspective*, proceedings of the Oxford Energy seminar, Mabro (ed.) R., Oxford Institute for Energy Studies, Oxford, 1986.

Kindleberger (ed.), C.P., *The International Corporation*, a symposium, The M.I.T. Press, Cambridge, Mass./London, 1971.

Kolde, E.J., *The Multinational Company*, Lexington Books, Toronto/London, 1974.

Kolko, J., *America and the Crisis of World Capitalism*, Beacon Press, Boston, 1974.

Koopmann, G., Matthies, K., Reszat, B., *Oil and the International Economy, Lessons from two Price Shocks*, Transaction Publishers, New Brunswick/Oxford, 1989.

Lacey, R., *The Kingdom*, Fontana/Collins, London, 1982.

Larson, H.M., Knowlton, E.H., Popple, C.S., *New Horizons 1927-1950, History of Standard Oil Company (New Jersey)*, Harper & Row, Publishers, New York, 1971.

Larson, H.M., The Rise of Big Business in the Oil Industry, in: *Oil's First Century*, Harvard Business School, Papers Given at the centennial Seminar on the History of the Petroleum Industry, November 13-14, 1959.

LeBel, P.G., *Energy Economics and Technology*, The Johns Hopkins University Press, Baltimore/London, 1982.

Lefeber, R., *Het Internationaal Energie-Agentschap (IEA), De overeenkomst inzake een Internationaal Energieprogramma (IEP), en het energierecht van de Europese Gemeenschappen (EG)*, unpublished, Leiden, 1986.

Lefeber, R., Linde, J.G. van der, Europese Integratie vergt een energie(k) beleid, in: *SEW* (6) juni 1987.

Levy, B., World oil marketing in transition, in: *International Organization*, vol. 36, no. 1, Winter 1982.

Linde, C. van der, Olie: de 'smeerolie' van het internationale marktsysteem, in: *Zicht op het Midden-Oosten*, Biegel (ed.), L.C., Boom/Venture Press, Amsterdam, 1985.

Linde, C. van der, Op zoek naar stabiliteit in de OPEC, in: *Op zoek naar stabiliteit in de Orient*, Linde, C. van der, Rooy, P. de (eds.), SDU, Den Haag, 1989.

Linde, J.G. van der, Lefeber, R., International Energy Agency captures the development of European Community Law, in: *Journal of World Trade*, vol. 22, no. 5, October 1988.

Luciani, G., *The Oil Companies and the Arab World*, Croom Helm/St. Martin's Press, London/New York, 1984.

Mabro, R., Bacon, R., Chadwick, M., Halliwell, M., Long, D., *The Market for North Sea Crude Oil*, O.I.E.S/ Oxford University Press, Oxford, 1986.

Mabro (ed.), R., *Natural Gas: an International Perspective*, proceedings of the Oxford Energy Seminar 1982-1985, O.I.E.S./Oxford University Press, Oxford, 1986.

Mabro (ed.), R., *OPEC and The World Oil Market, the genesis of the 1986 price crisis*, O.I.E.S./Oxford University Press, Oxford, 1986.

Mabro (ed.), R., *The 1986 Oil Price Crisis, Economic Effects and Policy Responses*, O.I.E.S./Oxford University Press, Oxford, 1988.

Mabro, R., The changing nature of the oil market and OPEC policies, in: *OPEC Review*, volume VI, no. 4, Winter 1982.

Mabro, R., *Political Dimensions of the Gulf Crisis*, GWO1, Oxford Institute for Energy Studies, Oxford, 1990.

Mabro, R., *OPEC's Production Policies. How do they Work? Why don't they Work?*, WPM12, Oxford Institute for Energy Studies, Oxford, 1989.

Magee, S.P., Multinational Corporations, the Industry Technology Cycle and Development, in: *Journal of World Trade Law*, vol. 11, no. 4, 1977.

Markusen, J.R., Melvin, J.R., *The Theory of International Trade*, Harper & Row, Publishers, New York, 1988.

Maull, H., *Olmacht, Ursachen, Perspektiven, Grenzen*, Europaische Verlaganstadlt, Frankfurt am Main, 1975.

MEED, *MEED Special report Saudi Arabia*, June 1987.

Meier, P.M., *Energy Planning in Developing Countries, an introduction to analytical methods*, Westview Press, Boulder/London, 1986.

Mikdashi, Z., *Transnational Oil, Issues, policies and perspectives*, Frances Pinter, Publishers, London, 1986.

Mohammad, Y.H., Mead, W.J., *World Oil Prices: Demand, Supply and Substitutes*, ICEED, Boulder, Colorado, 1990.

Molle, W., Wever, E., *Oil refineries and Petrochemical Industries in Western Europe, Buoyant Past, Uncertain Future*, Edward Elgar, Aldershot, 1984.

Molle, W., Oil Refineries and Petrochemical Industries: moving into maturity, in: *The Structure of the European Industry*, de Jong, H.W. (ed.), second edition, Kluwer Academic Publishers, Dordrecht, 1988.

Monsen, R.J., Walters, K.D., *Nationalized Companies, A Threat to American Business*, McGraw-Hill, New York, 1983.

Moran, T.H., *Oil prices and the future of OPEC, The political economy of tension and stability in the Organization of Petroleum Exporting Countries*, Resources for the Future, Research Paper R-8, Washington, D.C., 1978.

Moran, T.H., Modelling OPEC Behavior: economic and political initiatives, in: *International Organization*, 35, 2, Spring 1981.

Moran, T.H., Managing an oligopoly of would-be sovereigns: the dynamics of joint control and self-control in the international industry, past, present, and future, in: *International Organization*, vol. 41, no. 4, Autumn 1987.

Morgan, T., Oil prices- a long-term perspective, in: *OPEC Review*, vol. XI, no. 4, Winter 1987.

Mosley, L., *Grof Spel, Strijd om de olie in het Midden-Oosten 1890-1974*, Manteau, Brussel/Den Haag, 1974.

Nelson, R.R., and Winter, S.G., *An Evolutionary Theory of Economic Change*, The Belknap Press of Harvard University Press, Cambridge (Mass.)/ London, 1982.

Nore, P., Turner, T., *Oil and Class Struggle*, Zed Press, London, 1980.

North, D.C., *Institutions, Institutional Change and Economic Performance*, Cambridge University Press, Cambridge, 1990.

Niering, F., Oil Policy after Reagan, in: *Petroleum Economist*, February 1990.

Oakley, A., *Schumpeter's Theory of Capitalist Motion, A Critical Exposition & Reassessment*, Edward Elgar, Aldershot, 1990.

Odell, P.R., *Oil and World Power*, Penguin Books, Harmondsworth, 1986.

Odell, P., Vallenilla, L., *The Pressures of Oil, a strategy for economic revival*, Harper & Row, Publishers, London/New york, 1978.

OECD, *Historical Statistics, 1960-1987*, OECD, Paris, 1989.

OECD, *Economic Outlook*, 43, June 1988.

O&GJ, *Oil & Gas Journal Databook 1985, 1987-1990 editions*, PenWell publishing Company, Tulsa, Oklahoma.

Olsen, M., *The Logic of Collective Action, Public Goods and the Theory of Groups*, Harvard University Press, Cambridge (Mass.), 1971.

Olsen, M., *The Rise and Decline of Nations, Economic Growth, Stagflation and Social Rigidities*, Yale University Press, New Haven, 1982.

OPEC, *Annual Report*, Vienna, 1980-1988.

OPEC, *OPEC Annual Statistical Bulletin*, Vienna, 1966-1988.

OPEC, *OPEC Basic Oil Industry Information*, 1983.

OPEC, *OPEC Facts and Figures*, 1982-1990.

OPEC, *OPEC Official Resolutions and Press Releases 1960-1983*, Pergamon Press, Oxford, 1984.

OPEC, *OPEC Official Resolutions and Press Releases, Supplement*, Vienna, 1987.

Pachauri, R.K., *The Political Economy of Global Energy*, The Johns Hopkins University Press, Baltimore/London, 1985.

Peirce, W.S., *The Economics of the Energy Industries*, Wadsworth Publishing Company, Belmont, California, 1986.

Petzinger, Jr., T., *Oil & Honor The Texaco-Pennzoil Wars, inside the $11 billion Battle for Getty Oil*, G.P. Putnam's Sons, New York, 1987.

PlanEcon, *Soviet and East European Energy Databank*, volume two, Washington, D.C., March 1986.

Porter, M.E., *The Competitive Advantage of Nations*, Macmillan, London/Basingstoke, 1990.

Putterman, L., *The Economic Nature of the Firm, a reader*, Cambridge University Press, Cambridge, 1989.

Randall, S.J., *United States Foreign Oil Policy; for Profits and Security, 1919-1948*, McGill-Queen's University Press, Kingston/Montreal, 1985.

Rasmusen, E., *Games and Information, An Introduction to Game Theory*, Basil Blackwell, Oxford, 1989.

Razavi, H., Samii, M.V., Speculative demand for oil, in: *OPEC Review*, vol. VII, no. 1, Spring 1983.

Robinson, J., *Yamani, the inside story*, Simon and Schuster, London, 1988.

Roncaglia, A., *The International Oil Market, a case of Trilateral Oligopoly*, Macmillan, Basingstoke/London, 1985.

Sampson, A., *The Seven Sisters*, Coronet Books, Sevenoaks, 1988.

Saudi Arabian Oil & Gas Directory 1977-1978, in: *L'Arabie Saoudite d'hier a demain*, Cahier 8-9-10-11-12-13, Centre de Recherches sur le Monde Arabe Contemporain, Universite Catholique de Louvain.

Sayigh, Y.A., Arab economic Strategy in a Changing World, in: *Third World Quarterly*, January 1982.

Schubert, A., *Erdöl: Die Macht des Mangels*, Verlag Klaus Wagenbach, Berlin, 1982.

Schurr, S.H., Netschert, B.C., *Energy in the American economy, 1850-1975 an economic study of its history and prospects*, Greenwood Publishers, Westport, 1977.

Scitovsky, T., *Welfare and Competition*, Stanford University, Palo Alto, 1971.

Seymour, A., *The World Refining System and the Oil Product Trade*, GWO2, Oxford Institute for Energy Studies, Oxford, 1990.

Seymour, I., *OPEC Instrument of Change*, The Macmillan Press Ltd., London/Basingstoke, 1980.

Shepherd, W.G., *The Economics of Industrial Organization*, Prentice-Hall International, Englewood Cliffs, 1990.

Shwadran, B., *The Middle East, Oil and the Great Powers*, Atlantic Press, London, 1955.
Sinclair, S., *The World Petroleum Industry, The Market for Petroleum and Petroleum Products in the 1980s*, Euromonitor Publications, London, 1984.
Skeet, I., *OPEC: Twenty-five years of prices and policies*, Cambridge University Press, Cambridge, 1988.
Skeet (ed.), I., *Paul Frankel: Common Carrier of Common Sense, A selection of his writings, 1946-1988*, O.I.E.S./Oxford University Press, Oxford, 1989.
Sodersten, B., *International Economics*, Macmillan, Basingstoke/London, 1980.
Spero, J.E., *The Politics of International Economic Relations*, St. Martin's Press, New York, 1977.
Stork, J., *Middle East Oil and the Energy Crisis*, Monthly Review Press, New York/London, 1975.
Stournaras, Y., *Are oil price movements perverse? A critical explanation of oil price levels 1950-1985*, Oxford Institute for Energy Studies, WPM 6, 1985.
Studie-comite van de Westeuropese steenkolenproducenten, *Kolen en Energiepolitiek*, Brussel, juli 1955.

Tahmassebi, H., Crude oil and product differentials: historical perspective and outlook, in: *OPEC Review*, vol. VIII, no. 3, Autumn 1984.
Tanzer, M., Zorn, S., *Energy Update, Oil in the Late Twentieth Century*, Monthly Review Press, New York, 1985.
Tanzer, M., *The Race for Resources; Continuing Struggles over Minerals and Fuels*, Heinemann, London, 1980.
Teichova, A., Levy-Leboyer, M., Nussbaum, H., *Multinational Enterprise in Historical Perspective*, Cambridge University Press, Cambridge, 1989.
Terzian, P., *OPEC: the inside story*, Zed Books Ltd., London, 1985.
The Business History Review, *Oil's First Century*, papers given at the Centennial Seminar on the History of the Petroleum Industry, Harvard Graduate School of Business Administration, Nov. 1959.
The Economist, *Tribes with Flags, a survey of the Arab east*, 6 Feb 1988.
Tirole, J., *The Theory of Industrial Organization*, The MIT Press, Cambridge, Massachusetts, 1989.
Touscoz, J., *Quelle politique energetique pour l'Europe?, propositions pour une relance*, Presses Universitaires de Grenoble, Grenoble, 1986.
Tugendhat, C., Hamilton, A., *Oil, the biggest business*, Eyre Methuen, London, 1975.
Turk, J., *Transitions are Dangerous: Oil and the World Economy 1990*, Oxford Institute for Energy Studies, Oxford, 1990.
Turner, L., *Oil Companies in the International system*, George Allen & Unwin, London, 1983.
Tussie, D., *The Less Developed Countries and the World Trading System, a challenge to the GATT*, Francis Pinter, London, 1987.

United Nations, *Alternative Arrangements for Petroleum development: A Guide for Government Policy-makers and Negotiators*, New York, 1982.
United Nations, Prepared by the Secretariat, Economic Commission for Europe, *The Price of Oil in Western Europe*, E/ECE/205, Geneva, March 1955.
United States Senate, *The International Petroleum Cartel*, Staff Report to the Federal Trade Commssion submitted to the subcommittee on Monopoly of the Select Committtee on small business, Washington, D.C., 22-08-1952.
United States Senate, Hearings before the subcommittee on antitrust and monopoly of the committee on the judiciary, *The Petroleum Industry, part 1, Economist's Views*, Ninety-first Congress, first session S. Res. 40, Washington, D.C., March/April 1969.

Vietor, R.H.K., *Energy policy in America since 1945, a study of business-government relations*, Cambridge University Press, Cambridge, 1984.

Wall, B.H., *Growth in a Changing Environment, a history of Standard Oil Company (New Jersey) and Exxon Corporation 1950-1975*, McGraw-Hill, New York, 1988.
Werner, E.G.G., *The Shell Group and the Current Energy Scene*, a talk in the Volkshochschule of public meetings, Dusseldorf, 25-10-1978.
Williamson, H.F., Andreano, R.L., Daum, R., Klose, G.C., *The American Petroleum Industry, The age of energy*

1899-1959, Northwestern University Press, Evanston, 1963.

Williamson, J., *The Open Economy and the World Economy, a textbook in international economics*, Basic Books, Inc. Publishers, New York, 1983.

World Bank, *World Development Report 1987 and 1988*, Oxford University Press for the Worldbank, New York/London.

Wyant, F.R., *The Unites States, OPEC, and Multinational Oil*, Lexington Books, Lexington, 1977.

Yergin, D., *The Prize, The Epic Quest for Oil, Money and Power*, Simon and Schuster, New York, 1991.

Index

Studies in Industrial Organization

Series Editors:
H. W. de Jong, *University of Amsterdam, The Netherlands*
W. G. Shepherd, *University of Massachusetts, Amherst, U.S.A.*

Publications:

1. H. W. de Jong (ed.): *The Structure of European Industry.* Revised edition, 1988: see below under Volume 8
2. M. Fennema: *International Networks of Banks and Industry* (1970–1976). 1982 ISBN 90-247-2620-4
3. P. Bianchi: *Public and Private Control in Mass Product Industry.* The Cement Industry Cases. 1982 ISBN 90-247-2603-4
4. W. Kingston: *The Political Economy of Innovation.* (1984) 1989 2nd printing ISBN 90-247-2621-2
5. J. Pelkmans: *Market Integration in the European Community.* 1984 ISBN Hb: 90-247-2978-5; Pb: 90-247-2988-2
6. H. W. de Jong and W. G. Shepherd (eds.): *Mainstreams in Industrial Organization.* 1986
   Book I: *Theory and International Aspects.* ISBN 90-247-3461-4
   Book II: *Policies: Antitrust, Deregulation and Industrial.* ISBN 90-247-3462-2
   Set ISBN Book I + II: 90-247-3363-4
7. S. Faltas: *Arms Markets and Armament Policy.* The Changing Structure of Naval Industries in Western Europe. 1986 ISBN 90-247-3406-1
8. H. W. de Jong (ed.): *The Structure of European Industry.* 2nd revised ed. (of Volume 1). 1988 ISBN Hb: 90-247-3689-7; Pb: ISBN 90-247-3690-0
9. I. L. O. Schmidt and J. B. Rittaler: *A Critical Evaluation of the Chicago School of Antitrust Analysis.* 1989 ISBN 90-247-3792-3
10. B. Carlsson (ed.): *Industrial Dynamics.* Technological, Organizational, and Structural Changes in Industries and Firms. 1989 ISBN 0-7923-9044-X
11. Z. J. Acs and D. B. Audretsch (eds.): *The Economics of Small Firms.* A European Challenge. 1990 ISBN 0-7923-0484-5
12. W. Kingston: *Innovation, Creativity and Law.* 1990 ISBN 0-7923-0348-2
13. B. Dankbaar, J. Groenewegen and H. Schenk (eds.): *Perspectives in Industrial Organization.* 1990 ISBN 0-7923-0814-X
14. P. de Wolf (ed.): *Competition in Europe.* Essays in Honour of Henk W. de Jong. 1991 ISBN 0-7923-1050-0
15. C. van der Linde: *Dynamic International Oil Markets.* Oil Market Developments and Structure 1860–1990. 1991 ISBN 0-7923-1478-6

Kluwer Academic Publishers – Dordrecht / Boston / London

GPSR Compliance

The European Union's (EU) General Product Safety Regulation (GPSR) is a set of rules that requires consumer products to be safe and our obligations to ensure this.

If you have any concerns about our products, you can contact us on ProductSafety@springernature.com

In case Publisher is established outside the EU, the EU authorized representative is:

Springer Nature Customer Service Center GmbH
Europaplatz 3
69115 Heidelberg, Germany

Batch number: 10823959

Printed by Printforce, the Netherlands